MW01710876

Social Statistics: A User-Friendly Approach

Steven P. Schacht
Department of Sociology and Criminal Justice
Gonzaga University
Spokane, WA

with the assistance of

Brad Stewart
Department of Sociology
St. Thomas University
St. Paul, MN

Allyn and Bacon

Boston London Toronto Sydney Tokyo Singapore

Vice President, Social Sciences: *Susan Badger*
Senior Series Editor: *Karen Hanson*
Editorial Assistant: *Sarah Dunbar*
Marketing Manager: *Joyce Nilsen*
Cover Administrator: *Linda Knowles*
Manufacturing Buyer: *Megan Cochran*
Editorial-Production Service: *Lawrence Feinberg*
Cover Designer: Suzanne Harbison

Library of Congress Cataloging-in-Publication Data
Schacht, Steven P.
Social statistics: a user-friendly approach / Steven P. Schacht with the assistance of Brad Stewart.
p. cm.
Includes index.
ISBN 0-205-14004-1
1. Social sciences—Statistical methods. 2. Mathematical statistics. I. Stewart, Brad. II. Title.
HA29.S33 1995
519.5—dc20 94-43560
CIP

Printed in the United States of America
10 9 8 7 6 5 4 3 2 1 99 98 97 96 95

Preface

When I began teaching statistics nearly ten years ago in graduate school, I did so with excitement and apprehension. At that time I eagerly sought out any opportunities that allowed me to teach, and as a first year Ph.D. student I was thrilled to be teaching my fifth different course as an independent instructor. On the other hand, however, from my own experiences as a student and experiences other instructors shared with me, I knew all to well that I would face a generally hostile student audience. That is, I could expect many of the students in a statistics class to have insufficient math backgrounds, to be experiencing fairly high levels of math/statistics anxiety, and, in general, to be in attendance only because, as one student put it, "Some academic jerk requires it for graduation." Moreover, unlike the previous subjects I taught that seemed to have an inexhaustible supply of supporting materials, there were virtually no materials available to assist me in preparing a statistics class. In sum, after successfully teaching numerous courses, I was now faced with a subject that was generating instructor anxiety.

In previous courses, like many instructors I knew, I often used newspaper cartoons whenever controversial subjects that might provoke student anxiety were addressed. This teaching technique always worked well for me. Since many cartoonists present social science topics in a humorous light, I felt cartoons could be used both as a way to lower student anxieties and as an actual framework to investigate social phenomena. From that very first cartoon I used in a class, student response has been overwhelmingly favorable. Through personal research and from written comments on my student evaluations, I have found that students not only report this to be a fun, novel teaching technique but often cartoon examples are also seen as more applicable to their own lives, easier to understand; overall, they do appear to reduce student math/statistics anxiety. For more detailed discussions of these findings, please read the reprints of the *Teaching Sociology* articles found in this text's accompanying instructor manual.

Although I used a textbook the first couple of times I taught statistics, these materials seemed to be largely written for instructor instead of student needs. As such, I found that most statistics textbooks are of little pedagogical value to the average student (this is especially true for social science students) and I have only used one twice in the 15+ times I have taught this class. (Please see accompanying *Teaching Sociology* article on textbooks in statistics.) In place of a textbook, I have created a rather extensive set of take-home exercises for my classes based on cartoons. Moreover, all of the quizzes and exams in my classes are based on cartoon examples, as are most classroom presentations. This textbook is quite simply an outgrowth of this material.

There are three distinct features about this textbook that differentiate it from others. First, although this textbook is a comprehensive survey of statistical techniques found in most texts, it is still written at a level that the average behavioral or social science student can understand. Second, using cartoons in a statistics textbook not only makes traditionally dry material fun, but in many cases it actually reduces student anxiety. Finally, as an outgrowth of the first two features, I believe that this is one of the first statistics textbooks written with the student explicitly in mind. As such, I hope this textbook and the pedagogical techniques it advocates assist in the creation of a successful learning environment for both the instructor and the students.

I thank Brad Stewart for his editorial assistance on Chapters 1–6 and truly wish he would have been able to see this project through to its completion. I also thank professors Dan Pence of Southern Utah University, Robin Franck of Southwestern College, Diane Pike of Augsburg College, Michael Lacy of Colorado State University, and Victoria Swigert of the College of the Holy Cross for their detailed and thoughtful reviews. Their comments were quite helpful in my attempt to make this a truly user-friendly text. Special thanks are due Professor Ken Berry of Colorado State University. Although he was not directly involved in this project, many of his ideas and techniques for teaching statistics are found in this text.

There are myths that surround an author's relationship with his or her publisher. One of these myths apparently is that publishing companies and their editors are sometimes difficult to work with. My experience with Allyn and Bacon, however, has been just the opposite; Lawrence Feinberg, Judy Fiske, and Sarah Dunbar have all been a pleasure to work with and contributed significantly to the overall quality of this text. Moreover, I sincerely thank Karen Hanson for her understanding the endless delays, listening to what must have sounded like very tired excuses, and in renegotiating the text's completion. Her faith in this project is literally what made its completion possible.

Finally, I thank my life companion, Anna Papageorge. Walking the pathways of life, we are sometimes fortunate enough to befriend individuals who make our journey far more meaningful. She is definitely one of them.

Contents

Chapter 1

Introduction

We are willing to bet that you and many of the students in your class may read this first sentence with a fair amount of disdain. After all, statistics is often viewed as one of the most unpleasant and, in some cases, difficult of all the classes that you are required to take in college. As such and reflected in Cartoon 1.1, perhaps many of you are also math atheists.

CARTOON 1.1

Having taught statistics for many years, we have found that these two adverse attitudes—unpleasant and most difficult subject—are a result of several misperceived notions that many students hold about statistics classes. Although not all students despise the thought of having to take a statistics class, and some, believe it or not, actually look forward to taking this class, through research and classroom observations we have found several common themes for why most students do not look forward to taking this class.

To begin with, many students in the social sciences often view themselves as incapable of undertaking anything but fairly simple mathematical operations. In fact, many students select a major in the social sciences because of the belief that they are inept at any mathematically oriented subjects. Correspondingly, because of this attitude, many students also experience math anxiety. Although discussed in greater detail below, students who experience math anxiety, not surprisingly, try to avoid any classes that will elicit this feeling.

Further, many students who have not taken a statistics class often question the utility of such a course to their "academic learning experience." That is, students often believe the only value in taking this course is that it is required to graduate. This general adversity to statistics is also a result of the widespread student attitude that statistics have very little applicability to other college courses, not to mention their careers or to life in general. Perhaps this explains why so many students postpone taking statistics until their senior year. Of course, this often causes students to have additional anxieties: "If I fail this class I won't graduate this term." In total, these sentiments often lead to the an overriding attitude that this course is going to be a potentially painful, basically worthless experience.

Why Learn Statistics?

While the rest of this textbook is written to dispel these negative *myths*, some preliminary arguments about the utility of statistics are warranted. If you agree with the adage that "there are lies, damn lies, and statistics," then you already have a very good idea of the potential value of taking this course (this paraphrased sentiment was initially put forth by Mark Twain). Regardless of your chosen career, you will constantly be presented with statistical information. For instance, your boss might want you to review a recent statistical report on the demographics of the county to estimate future social service demand. Or, alternatively, the instructor for your theory class may require statistical evidence to support a research paper you are writing for her or his class.

Whenever those who are not knowledgeable in statistics come across any statistically based information, they have to either unquestionably accept what others have concluded from it (assuming they even understand this) or find someone to interpret it for them. The latter can be quite expensive. Either way, you run the risk of being deceived or even outright lied to because of your statistical ignorance. As summarized by another paraphrased saying: Figures often lie and liars often figure. Such ignorance can be potentially costly to both your employer and your career. Quite simply, being knowledgeable in statistics is an asset to your career aspirations.

Outside of our workplaces, moreover, we are bombarded daily by statistical information. Like it or not, we live in a statistical society. Everything from your car insurance rates, opinion polls, weather reports, crime rates, to consumer information is based on statistics. Being conversant in elementary statistics makes each of us better consumers and less likely to be "ripped off" in our purchases. Additionally, and perhaps more importantly, statistical knowledge makes each of us better informed citizens. This allows us to view more rationally the social issues that confront us.

In summary and at the risk of sounding redundant, like it or not, we all reside in a statistical world. As such, at varying degrees of understanding, we are also all users of statistics. Thus, regardless of one's chosen major or career, having a basic, rudimentary understanding of statistics is an asset to anyone who hopes to get ahead or even stay even in today's society. With these preliminary thoughts in mind, we now turn to the goals of this text and how they will ultimately make the task of becoming statistically informed a much easier accomplishment than you probably ever thought possible.

Goals of This Textbook

Prior research by us and others has found that social science students often dislike statistics because they suffer from math/statistics anxiety. After all, as reflected in Cartoon 1.2, certain majors (e.g., humanities and social sciences) are supposed to minimize the number of math-oriented classes one has to take. Math/statistics anxiety usually takes the form of students' fearing that they will do poorly or fail a given statistics class because they are incapable of doing the math required for it. This often becomes self-fulfilling.

CARTOON 1.2

FRANK AND ERNEST reprinted by permission of UFS, Inc.

Students suffering from this type of anxiety, regardless of how mathematically inclined they are, almost always experience more difficulties learning the material than those who do not have this affliction. Moreover, people who suffer from high levels of math anxiety often try to keep their feelings secret. Unfortunately, keeping this secret precludes a given student from seeking help from his or her classmates and teachers out of fear of appearing stupid. As many of you have probably guessed, extreme cases of math/statistics anxiety that are not recognized and dealt with in some meaningful manner can often result in student failure.

Not surprisingly, then, often students' math anxiety also frustrates instructors. Some instructors' inability to recognize this very real classroom problem leads them to conclude wrongly that their students are lazy and unprepared. This false impression of the instructors combined with students' real anxieties (albeit these, too, are almost always unfounded) becomes a vicious cycle where a negative learning environment becomes an unfortunate but inevitable outcome. Although the problem of math/statistics anxiety is often quite real for both the student and the instructor, fortunately it is always unnecessary.

Therefore, the first goal of this textbook is to lessen each student's statistical anxiety. As noted, this textbook approaches statistics far differently than any other that your instructor could have selected. Not only is this text written in a manner that addresses students' anxieties, it utilizes a specific medium we have found successful in dealing with this problem: cartoons.

Humor is an excellent way to deal with anxiety. For instance, have you ever noticed that people often laugh when they are nervous or uncertain about something? This laughter serves to reduce tension and anxiety. This same principle can be applied to math anxiety. Specifically, we use cartoons, such as the two already presented, to make humorous statistical presentations that are often perceived as boring, intimidating, and difficult. We have found that being able to laugh at the given material with a cartoon almost always reduces the amount of math/statistics anxieties present. Reducing anxieties ultimately allows students to more easily learn and retain statistical principles and operations that are in reality—at least at the introductory level—not that difficult, not to mention that approaching a subject such as statistics using humor can actually make learning the material fun.

Further, in an attempt to lessen remaining math anxieties you might have, it should be noted that since this is an introductory statistics course, a strong math background is not required or assumed. If you are comfortable with the basic mathematical operations of addition, subtraction, multiplication, division, how to square a number, and how to obtain a square root—all operations found on the most basic of calculators—you should not have any problems with the math found in this textbook. Moreover, all of these basic operations are comprehensively and contextually reviewed in Chapter 2.

In sum, with cartoons serving as a common medium, this textbook is concerned with offering each student an introduction to statistics. Fundamental ideas and concepts (formulas) that most statistical analysis is based upon are presented in a basic, systematic, and—hopefully—humorous manner. These presentations are structured so that each student will gain an understanding of the practical, applied nature of statistics. Almost every presentation begins with a non-intimidating, often absurd, cartoon example typically followed by a real-life example. In other words, the overall goals of this text upon its completion are twofold: (1) to lessen each student's potential anxieties about statistics and (2) to demonstrate the utility of having a working knowledge about statistics.

All of this is not to say this is a watered down, overly simplistic textbook. It is not. All concepts found in similar introductory statistics textbooks are comprehensively covered in our text. Unlike every other statistics textbook that we are aware of, however, this textbook is truly and explicitly written for you the student. Since assuredly one of the reasons your instructor selected this text is because of its student emphasis, we are guessing that you will not find this course to be the painful, basically worthless experience you might have expected. Rather, we are willing to wager (with the odds definitely in our favor) that you will find statistics are a meaningful, sometimes funny way of describing and understanding the world we reside in.

WHAT ARE STATISTICS?

The foremost goals of the social sciences are to *describe* and *explain* different types of social phenomena. That is, social scientists ultimately strive to understand how social characteristics relate to and potentially affect each other. Social characteristics are also referred to as variables. As the word implies, variables are characteristics that vary (*vari*ables) in some measurable way from unit to unit.

When relationships between variables are explored, one characteristic (called the *independent variable*) is often viewed as the cause of change in another characteristic (called the *dependent variable*). While independent and dependent variables are extensively discussed in later chapters, at this point it is important to realize that whenever we want to explain and understand social phenomena we almost always do so in terms of relationships and causes. An independent variable (or variables) quite simply is seen as bringing about changes in a dependent variable (or variables).

In other words, independent variables are seen as determinants of dependent variables. Examples of simple independent/dependent variable relationships are (a) years of education seen as a determinant of income and (b) one's gender (and sexist attitudes) seen as determinant of one's major. While there is an infinite array of social variables and possible relationships among them, when social scientists explore such relationships they almost always do so in terms of independent and dependent variables.

Additionally, variables (social characteristics) in the social sciences are seen as occurring in one of three basic forms. Using Cartoon 1.1 and Calvin as a backdrop, these three types of characteristics are (1) *personal characteristics*, such as gender (he is a male) or religious background (a math atheist), (2) *attitudes*, such as degrees of liking or disliking mathematical subjects (he has an apparent animosity towards math), and (3) personal *behaviors*, the way a person acts (apparently he also avoids mathematically oriented topics).

Then again, since the focus of any social scientist's research is that which is *social*, we are obviously not interested in just how one individual thinks and acts, but rather how a multitude of different types of people in general think and behave. For instance, it is funny and somewhat interesting to note that Calvin is a math atheist who doesn't like this subject and probably avoids it at all costs. But beyond Calvin and his typically tortured existence, reflected in this cartoon, who cares?

As a social scientist, however, what if we knew that numerous people in our society (including members of your class) were math atheists who avoided this topic at all costs? Not only would this be of great interest to us but to learn anything further we would need some meaningful way to describe and hopefully understand this social phenomenon. This is what statistics enable us to do. While statistics as a scientific tool allows us to do many things, at its most fundamental level statistics allows us to summarize quantitatively and to generalize social phenomena so that we may better understand their occurrence.

As reflected by Calvin, often social phenomena are indicative of a social problem: math anxiety. Statistics can be used to describe and better understand nearly any social problem; math anxiety is detrimental to student learning while the use of humor appears to reduce such anxieties. Moreover, new-found understandings almost inevitably lead to innovative ways of dealing with the social problem to be proposed and tested: a cartoon statistics textbook.

All of this is not to say that statistics is the only meaningful, or even best, way for social scientists to gain new insights and understandings of social phenomena. Studies using statistical analyses, however, are the most widely used and accepted form of research undertaken in the social sciences.

Thus far, the term "statistics" has been used rather loosely without a formal definition being offered. Figure 1.1 helps explain what statistics are and are not. Often it is easier to define what something is by first recognizing what it is not. As such, we will first define what populations and parameters are, which will in turn enable us to gain a more complete understanding of statistics.

Populations

To begin with and as reflected in Figure 1.1, statistics are obviously not populations. A population can be defined as nearly anything. Examples of things that are often defined as populations are everyone in the United States, all the students at a given university, or a classroom of students. In practice, a social scientist's research question usually dictates the population s/he is studying. For instance, if you were doing a study on the level of students' math/statistics anxiety in a given statistics class, your population is obviously not students currently enrolled in an anthropology course. Rather, your population is students currently enrolled in the statistics class of interest. Students from one statistics class or several could be defined as the population, but

FIGURE 1.1 Population Parameters and Sample Statistics

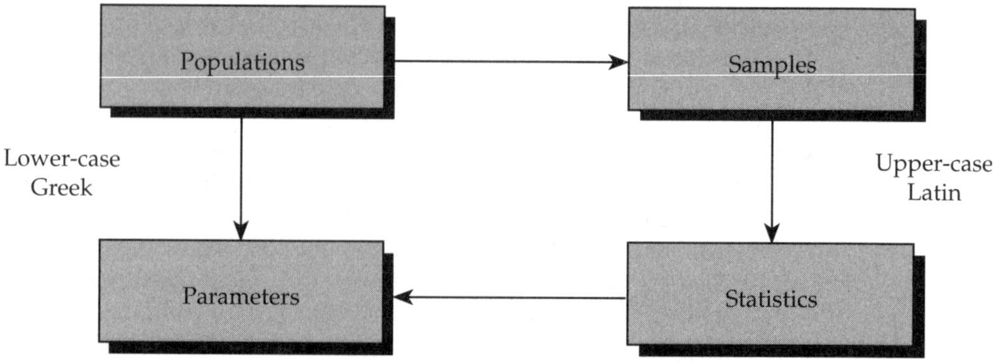

only these types of students could be considered since that is what your research question dictates.

Once a population has been defined, then, every unit within it is treated in an equal manner. Referring back to the statistics class example, if there were 30 students in the class and you defined them as your population, then all 30 students' math/statistics anxiety levels would have to be measured. Although discussed in greater detail in just a moment, when a population is used, answers to accompanying mathematical procedures (such as an average attention span) are called parameters. If, however, you missed just one student in the present example and only measured 29 subjects' anxiety level, you would no longer have a population parameter, but rather a sample statistic. In other words, measuring anything less than the total of your defined population is no longer a population of values; it is a sample statistic. Before samples and statistics are explored, however, the math associated with populations must first be discussed.

Parameters

Referring back to Figure 1.1 and as just noted, the mathematics associated with populations are called parameters. Potential examples of parameters are averages, proportions, and percentages. The key to understanding a parameter is that every unit within the defined population must be utilized. Once again, if we defined our population to be 30 students in the hypothetical statistics course of interest, measured the math/statistics anxiety levels of all 30 students, and then calculated an average, the resultant answer would be called a parameter. The mathematical notation typically associated with population parameters are lower-case Greek letters. Mathematical notation is defined and discussed in detail in the next chapter.

Additionally, population parameters are considered *descriptive*. That is, since every element within the given population is considered, any resultant summary measures—parameters—are correspondingly used to *describe* everyone. For instance, building upon on our previous example, let's say that on the first day of class we measured each student's (all 30) level of math/statistics anxiety and found the class average to be 8 (on a scale of 1 to 10; averages/means are discussed in detail in a subsequent chapter). Since every element (student) was mathematically considered, then the reported value of 8 is an exact, single summary measure—description—of the entire population.

While statistics can also be used descriptively, more importantly they enable us to make what are called *inferences* and to *test hypotheses*. To understand what these two newly introduced terms mean, we must finally turn to the task of defining exactly what samples and statistics are.

Sample

As already noted, a sample is defined as anything less than the entire population. Mathematically, a sample is expressed as $N-1$, where N equals the total number of elements within a given population. Almost always, however, a sample is a much smaller

FIGURE 1.2 Drawing a Sample from a Population

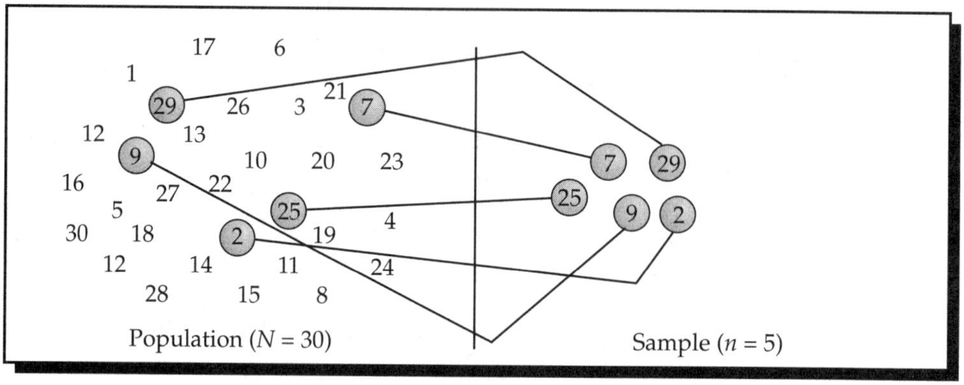

portion of a larger population. Using the statistics class example (N = 30), Figure 1.2 graphically demonstrates how a random sample can be drawn from a population. The term "random" and its importance are discussed in a moment.

There are three primary reasons why social scientists usually conduct research using a sample instead of a population. First of all, gathering data from even a small population is extremely *expensive*. In the social sciences, however, we are often interested in large populations. For instance, let's say we are interested in the average income of a hypothetical community of 100,000 households. To obtain a parameter average income value we would have to interview every household—all 100,000 of them. Actually doing this would be extremely expensive, and quite unnecessary considering that a sample can give us a precise estimate of any parameter value. In fact, a sample of 400 households and their average incomes would give us a fairly accurate estimate of all 100,000 households for a fraction of the cost of obtaining the parameter average.

To obtain complete information from even small populations is not only expensive but also an extremely *time*-consuming process. That is, a second reason why samples are often used is that it simply takes too long to completely gather data from a population. Using our example of 100,000 households again and assuming we had the resources, it would take months with dozens of interviewers to obtain every household's income. Moreover, over a period of months each household's income could drastically change due to inflation, unemployment, or other unforeseen circumstances. This could potentially render any final parameter obtained meaningless since it may no longer be true.

The problem of time constraints is even further compounded with attitudinal surveys. For instance, let's say we are interested in approval ratings of the current President. Obviously, however, sentiments on nearly any issue vary over time; with topics such as the President's approval ratings, they sometimes change literally overnight. In other words, a position initially expressed may no longer be true days later. Exactly because of time constraints and the adverse effect they can have on one's findings, most political polls (samples) are conducted overnight or within a 24-hour period.

In sum, since social scientists' research subjects are actual people who are constantly changing, most data collection efforts must be completed in a timely manner. A sample is a far more efficient way (sometimes it is the only way) of rapidly collecting information on a population of people.

This leads us to the third related reason for samples; it often is *impossible* to gather a population of information. Even in situations where we might have the money and guard against the possibility of changing sentiments, we could not expect to reach all potential respondents. People might be at work, out of town on business or vacation, in the hospital, or simply unavailable. Moreover, these potential availability problems don't even take into account people who might refuse to participate in our study. In other words, when people can't be located or refuse to participate, we are left with a sample by default ($N - 1$). In total, not only is it usually impractical due to time and financial constraints to interview everyone in a defined population, but it is often impossible.

This is equally true in other fields. When you have a blood test at the doctor's office, would you rather have a sample or a population drawn to be analyzed? In industry, companies often make claims about the longevity of their products. For instance, GE might make the claim that their light bulb has an average life of 400 hours or their batteries last an average of 40 hours. How do they arrive at these figures? They simply test a sample of the given product. To do otherwise would require that all the units were used—such as a population of burned-out light bulbs—and this would leave the company no products to sell.

Then again, many of these same considerations for why samples are drawn are applicable to life in general. When you come across milk that you suspect has spoiled or foods that you think you might dislike, do you consume all of the given substance or simply take a *sample*? If you come to college without a selected major, should you take all the courses offered at the given school to make this decision or just a few introductory courses in field areas that potentially interest you? A sample is not only the most logical choice, but it is by far the most feasible way to answer any of the above questions. In other words, we are all already users of samples.

In sum, samples are not only a practical way to collect data, personally and scientifically, but sometimes they are the only way in which information from a large population can be collected. With this in mind, we are now ready to discuss what the math associated with samples (statistics) enable us to do: make *inferences* about population parameters and test hypotheses.

Statistics

Finally we arrive at a basic definition for statistics: quite simply, the math associated with samples. Statistics are obviously much more than this, but at their most basic level of meaning they are "sample math." The mathematical notation typically used to represent statistics are upper-case Latin letters.

Before proceeding to a fuller discussion of statistics and what they enable us to do, two key assumptions concerning sample statistics must be discussed: *randomness* and *representativeness*. For the purposes of this text, randomness means that once a given population has been defined, typically in the form of a list, then every unit on it has an equal chance of being selected. This can be accomplished two basic ways: (1) We can systematically take every "*n*th" unit from a population list or (2) we can randomly select the units with a table of random numbers or a calculator that randomly generates numbers. Either way, the underlying assumption of randomness is that each sampling unit has an equal chance of being selected. In Figure 1.2, a sample of 5 drawn from a population 30 was drawn—the numbers selected—using a table of random numbers.

If a sample is randomly generated, then it is also usually thought to be representative. The assumption of representativeness simply means the sample drawn reflects—represents—the larger population. Although we can never absolutely prove it, randomly selected samples are always generated with the idea that they are representative of the larger population from which they were drawn. If a sample is not randomly generated, then statistically speaking we cannot assume it to be representative of the larger population from which it was selected.

When a random sample is drawn that is considered representative, resultant statistics are not only used to describe a sample but are also used to make inferences and hypothesis test. *Inferences* quite simply are *estimates* of population parameters. Thus, a given sample average is the best and most logical estimate of the true parameter average. Comparisons of averages representing different groups within the sample are also often undertaken; this is a form of *testing hypotheses*. While both of these procedures are discussed in detail in subsequent chapters, estimating population parameters and hypothesis testing are the two basic ways that statistics are utilized.

Descriptive Versus Inferential Statistics

Many statisticians find it useful to also differentiate between descriptive and inferential statistics. The distinction between these two terms is really an expansion of our discussion so far. To begin with and as already noted, all population parameters are descriptive. That is, parameters describe certain characteristics of a population. These descriptions can be averages, proportions, or numerous other mathematical calculations of a variable. They can also describe the relationship between two or more variables. However, the resulting findings cannot be generalized beyond the population from which they were generated. Thus, population parameters are always descriptive.

Statistics are also used in a similar manner. Correspondingly, they are called *descriptive statistics*. Identical to parameters, descriptive statistics are used to summarize a given variable. Also, like parameters, they simply summarize the values from which they are generated and are not used to make generalizations. The only difference between the two is that descriptive statistics are calculated using a sample while parameters are derived from a population of values. While this differentiation has important implications for how certain formulas are actually calculated, parameters and descriptive statistics are used to the exact same end: to *describe* a set of numbers.

Inferential statistics, however, are quite simply the previously outlined estimates of population parameters and hypothesis testing. As you have probably already concluded, they are also extremely important to social science research. Although there are other equally meaningful ways that social scientists gather data, only inferential statistics enable us to make scientific generalizations about large groups of people. Using a random sample that is usually just a fraction of the larger population from which it was drawn, inferential statistics enable us to make relatively inexpensive and timely generalizations. Moreover, as previously noted, using a sample is sometimes the only way in which information about a population can be gathered.

In total, inferential statistics are important, powerful scientific tools available to anyone who hopes to better understand the social world we all reside in. We believe that individuals who understand how statistics are calculated and what they represent, correspondingly, are often in a position to better understand our social world. While it is doubtful that we could convince Calvin of the value of understanding mathematics and statistics, especially considering that he is generally a strong proponent of an "ignorance-is-bliss" mentality (just ask Hobbes), the above hopefully has given you a preliminary indication why such knowledge might be worth having. Of course, this is an "inference" on our part.

Box 1

Often you will hear sports announcers state something like the following: "Kirby Puckett of the Minnesota Twins on the way to winning the 1989 American League batting title had career-high 'statistics'." Similarly, most newspapers and sports magazines also report the accomplishments of athletes as statistics: individual statistics, team statistics, pitching statistics, or quarterback statistics, to mention a few. What they are actually reporting are parameters. Once again, statistics are the math associated with samples while parameters are the math associated with populations. Thus, when a quarterback is reported as having a 55% completion rate, this value is calculated by taking all successful passes and dividing this value by all attempted passes. Since "all" attempts represent a population of values, the calculated completion rate is a parameter. To actually be a statistic we would have to take a sample of all attempts.

Then again, which sounds better, the San Fransisco 49ers team statistics or parameters? Perhaps this explains the continued misuse of the term "statistics."

Chapter Summary and Conclusions

This chapter offered some basic definitions of what statistics are and are not. Although statistics are obviously not populations or parameters, they are directly related to these terms. Populations can be defined as nearly anything. Every resident of your state or every student in the statistics class you are attending are both potential examples of populations. The math associated with populations are called parameters. Parameters are always descriptive.

Alternatively, statistics are quite simply the math associated with samples. Samples instead of populations are often utilized in the social sciences because they are cost effective, timely, and practical. Statistics are either descriptive or inferential. Descriptive statistics are used to summarize variables and variable relationships. Inferential statistics, however, are used to scientifically estimate population parameters and test hypotheses. As such, inferential statistics are vital tools that social scientists use.

Finally, this chapter offered some preliminary reasons why learning how to use and understand statistics might be of great value to you. We also recognized that for students to make this realization, their math/statistics anxieties must be not only recognized but also dealt with in some meaningful manner. Your instructor by selecting this textbook must also think this is an important issue in successfully teaching statistics. Using cartoons, the rest of this textbook is devoted to alleviating student anxieties so that we can demonstrate how statistics can be used as a meaningful, sometimes funny way of better understanding social reality.

Key Terms To Remember

Descriptive Statistics	Parameters
Estimates of Parameters	Populations
Hypothesis Testing	Samples
Inferential Statistics	Statistics
Independent and Dependent Variables	

Practice Exercises

1. If we talk about a baseball player's batting average and view it in terms of all trips to the plate, then this is a __________, not a __________.
2. Three reasons were offered in this chapter for why social scientists typically utilize a sample instead of a population.
   They are
   a. ______________
   b. ______________
   c. ______________

3. Characteristics measured in the social sciences, such as age or income, that vary from person to person or unit to unit are called ________.
4. Variables in the social sciences were proposed as occurring in one of three basic forms. These are
   a. ____________
   b. ____________
   c. ____________
5. Through research and personal observations, we have found that using humor, specifically cartoons, is an excellent way to reduce students' math/statistics anxiety. Thus, our two variables of research interest are humor and math/statistics anxiety. As such, ________ is seen as the independent variable while ________ is seen as the dependent variable.
6. As evidenced in Cartoon 1.1, Calvin apparently experiences such high levels of math anxiety that he has become a math atheist. Correspondingly, he tries to avoid any mathematically oriented topics. In this instance, math avoidance is seen as the ____ variable while math anxiety is seen as the ________ variable.
7. Statistics that are used to summarize a given variable are called ________ statistics.
8. Statistics that are used to estimate population parameters or to test hypotheses are called ________ statistics.
9. Conversely, inferential statistics enable us to do two basic things. These are
   a. ____________
   b. ____________

Chapter 2

Basic Mathematical Concepts

The last chapter offered an introduction to both the applicability and utility of statistics. Before proceeding to actual statistical analyses, however, a basic discussion of mathematical concepts must be presented. Although some of you who are more mathematically versed may find this chapter to be an overly simplistic review, we suspect that most of you will find it to be quite useful.

Regardless of one's math background, since the rest of the text simply builds upon the following presentation, all students should thoroughly familiarize themselves with it. To assist you, this chapter is written in such a manner that even those lacking math backgrounds should find it quite simple to follow and understand. Since the rest of this text is nothing more than variations of the mathematical concepts explored in this chapter, you should take this to be a quite promising sign—a strong indication that you will successfully survive this course.

The chapter ends with a discussion of levels of data measurement found in statistical analyses. Although this material is also quite easy to understand, it is imperative that you become familiar with it, too, since levels of measurement are determinants of what types of statistical analyses are subsequently utilized.

Math As a Recipe

All of us, at one time or another, have prepared and cooked food. In learning to do so, most of us followed a recipe. A recipe, offered by another person or found on a card/package, is quite simply the cooking directions for a given dish. Failure to follow such directions, as many of us can sadly attest, leads to ruined food and sometimes to an empty stomach.

To a certain extent, mathematical formulas can be likened to recipes. That is, the elements used to give instructions with a recipe are quite similar to those used with a mathematical formula. These elements are ingredients, amounts, operations, and order. Also, similar to using a recipe, failure to follow the correct directions of a given formula, not surprisingly, causes the resultant answer to be wrong. This becomes exceedingly clear as we explore the following analogy.

Ingredients

The ingredients used in recipes are quite simply the materials to be prepared and cooked. These include things such as flower, sugar, yeast, salt, and so forth. There are also ingredients in statistical formulas. These are, as discussed Chapter 1, *variables*. Once again, variables are any measurable social characteristics (e.g., your age, gender, or a recent exam score). Mathematically, variables are often represented by letters; statisticians often use the letters of X, Y, or Z. In a very real sense, variables measuring distinct elements are like the flavors of different ingredients in a dish. Moreover, when different variables are considered together, like a dish, a whole new element/flavor often emerges.

Amounts

Recipes also include the amounts of different ingredients to be used. For instance, for a given recipe one might use the amounts of two cups of sugar, one teaspoon of baking soda, and eight ounces of milk. In other words, amounts are *numerical* values of different ingredients: *two* cups, *one* teaspoon, and *eight* ounces. This is also true in mathematical/statistical formulas. For example, a person can be 30 years old and make $30,000 a year. The values of 30 and 30,000 are, respectively, amounts of the variables of *age* and *yearly income*.

Operations

The completion of certain operations are additionally required with recipes. For instance, in making a pizza crust from scratch one is required to *knead* the dough and *spread* it on a pan. Specific operations are also required in math/statistical formulas; these are the familiar operations of *addition*, *subtraction*, *multiplication*, and *division*.

Order

To be a successful cook, one must not only use the specified ingredients, in certain amounts, and perform the right operations, but all of the operations must be done in a specific order. Using the pizza crust example again, one must first *knead* (1) the dough before it can be *spread* (2) onto a pan to be subsequently *baked* (3). If these operations are not completed in the correct order (such as baking the dough before it is spread onto the pan), one ends up with a culinary disaster instead of a delight.

A certain order to the operations must also be followed on mathematical/statistical formulas. As many of you are aware, these are called the *rules of algebraic order*.

TABLE 2.1 Rules of Algebraic Order

Algebraic Order
1. ( )
2. exponentiation
3. × and/or ÷
4. + and/or −

Table 2.1 summarizes the specific order in which mathematical operations must be completed. Stating these in English, first all operations in parentheses (e.g., $X - Y$) are completed. Second, all operations of exponentiation (e.g., the number 3 squared) are done. Third, after these first two operations are finished, in this order, all operations of multiplication and/or division are undertaken. The final mathematical operations completed are addition and/or subtraction.

Since the rules of algebraic order are not only applied to the material found on the remaining pages of this chapter but the rest of text, if you are not completely familiar with them, we strongly urge you to either mark this page and/or write them down somewhere prominently in your notes. After all, like recipes and cooking, when using mathematical formulas, if one fails to follow all the operations in the specific order asked, any results will also be a disaster—*the wrong answer*.

SUMMATION NOTATION

Recipe instructions often make use of abbreviations and special symbols such as cups (c.), tablespoons (Tbsp.), or teaspoons (tsp.). Mathematical formulas, too, use special symbols and abbreviations as instructions. While many of these are the familiar symbols used to represent basic mathematical operations (×, ÷, +, and −), some symbols used in statistical formulas may appear quite foreign to many of you. More specifically, most statistical formulas use specialized symbols to represent an additional mathematical operation called summation notation. Actually, this procedure is a variant of a mathematical operation you are already quite familiar with: addition. Regardless, to best demonstrate what this newly introduced mathematical symbol instructs us to do, we need finally to undertake some mathematical calculations.

To assist in this discussion, we offer Cartoon 2.1 and its corresponding data sets found in Table 2.2. The first *data set* (group of numbers found under the X column heading) is measurements of the variable (level of desire for a "bush" in the White House) for 10 different dogs. More specifically, on a scale of 1 (low) to 10 (high), this first set of scores represents 10 dogs' self-reported levels of desire—*do not* to *do*—for a "bush" in the White House. The second set of numbers, found under the Y, represents the number of times one was used from 1989 to 1993. In other words, the second data

CARTOON 2.1

FRANK AND ERNEST reprinted by permission of UFS, Inc.

set represents the variable "number of times a "bush" was used" by the same 10 dogs whose level of desire for a "bush" in the White House was also measured.

We can also view these two data sets as representing an independent and dependent variable. In fact, the letters X and Y are traditionally used by statisticians to represent, respectively, independent (X) and dependent (Y) variables. (If the terms "independent variable" and/or "dependent variable" are at all confusing to you, please refer back to Chapter 1 for a more detailed discussion.) For the present example, then, the independent variable is self-reported level of desire for a "bush" in the White House (X) whereas the dependent variable is number of times one was used (Y). As such, desire for a "bush" is conceptualized (seen) as a determinant (a cause) of the usage. When the two data sets are considered together, as desire for a "bush" levels increase, the number of times a dog uses a "bush" in the White House correspondingly increases.

TABLE 2.2 Bush-in-the-White House Data Sets

	Desire a "bush" in the White House (X) Independent Variable	*Number of Times a "bush" was Used (Y) Dependent Variable*
Dog	(X)	(Y)
1	1	2
2	2	2
3	3	3
4	4	5
5	4	7
6	5	9
7	5	11
8	7	11
9	8	12
10	9	13

Box 2

We should note that the above data sets and most of the data sets that follow in this text are simply made up as hopefully fun examples. This, however, is not to say that the examples given are not applicable to real-life situations. To the contrary, every example found in this text is potentially applicable to some real-life circumstance. For instance, while the present cartoon example may seem somewhat absurd, as we have already noted, math phobia (anxiety) can adversely effect an individual's performance when answering a mathematical problem. Thus, alternatively viewing the values reported for the independent variable (*X*) in Table 2.2 as level of math/statistics anxiety and the dependent variable (*Y*) values as "incorrect answers" on a recent 25-question statistics exam, the data sets would then potentially reflect a real-life situation; math/statistics anxiety adversely effects students' math exam scores.

In other words, although they often appear seemingly absurd, the examples found in this text are still realistic insofar as one's imagination makes them such. We believe that using one's imagination (and humor) not only makes the reasoning behind statistics easier to grasp, it also makes learning the material more fun.

With this in mind, let's say we are given the following simple mathematical formula called the "sum of X." (As we will see, this is actually a sub-part of several larger statistical formulas. For this and the next few examples, however, more simply we will refer to these formula sub-parts as "formulas.")

Sum of X: ΣX

You are already familiar with the *X* in this formula; it informs us that we are dealing with an independent variable. The symbol that precedes X, Σ, is called sigma. Sigma, in mathematical terms, denotes summation. Or, stated slightly differently, sigma instructs us to sum a set of numbers. This is an example of summation notation. Applied to the "desire" data set (column 1 of Table 2.2), the sum of *X* instructs us to add together all observed *X* values. Done below, the sum of *X* is 48.

$$1 + 2 + 3 + 4 + 4 + 5 + 5 + 7 + 8 + 9 = 48$$

If, alternatively, the ΣY is asked for, it is calculated by simply taking all *Y* values and summing them together; with the above data set, this is $2 + 2 + 3 + 5 + 7 + 9 + 11 + 11 + 12 + 13 = 75$.

In slightly more complicated statistical formulas that involve the use of summation notation, the previously discussed rules of algebraic order must also be applied. Unfortunately, however, we have found that some students often become confused when summation notation is used simultaneously with the rules of algebraic order. As such, and considering that most statistical formulas use both of these statistical procedures, we suggest that you pay special attention to the discussion that follows.

Found below are two formulas that appear to be similar, but are mathematically quite different. That is, while both formulas instruct us to sum and square values of the independent variable (X), because of the rules of algebraic order, these operations are completed in basically opposite sequences. As a result, when these two formulas are applied to the exact same set of numbers, two very different answers are obtained.

Sum of All Squared X Values	Sum of X Values Squared
ΣX^2	$(\Sigma X)^2$

Thus, the first formula, appropriately labeled "Sum of All Squared X Values," tells us to separately square each of the X values (each observed value is multiplied by itself, e.g., $2 \times 2 = 4$ or $8 \times 8 = 64$), and then add them together. The reason for this specific order is that, once again, *exponentiation* precedes *addition*. Applying this formula to the first "bush"-in-the-White-House data set, then, we first individually square each X value. This is accomplished below.

$1 \times 1 = 1$;	$2 \times 2 = 4$;	$3 \times 3 = 9$;	$4 \times 4 = 16$;	$4 \times 4 = 16$;
$5 \times 5 = 25$;	$5 \times 5 = 25$;	$7 \times 7 = 49$;	$8 \times 8 = 64$;	$9 \times 9 = 81$

Next, as demonstrated below, to derive the final answer we simply add these 10 separate answers together. Thus, the sum of all squared X values is 290.

$$1 + 4 + 9 + 16 + 16 + 25 + 25 + 49 + 64 + 81 = 290$$

$$\text{Thus the } \Sigma X^2 \text{ equals } \boxed{290}$$

In the second formula, labeled "Sum of X Values Squared," we are instructed to add together all the X values, and then square this single value to get the final answer. Once again, the reason for this, as dictated by the rules of algebraic order, is that mathematical operations within *parentheses* (even when adding) always precede *exponentiation*.

When this second formula is applied to the same data set just used, as expected, we get a very different answer. Done below, we find that the sum of X values squared equals 2304 (versus 290).

$$1 + 2 + 3 + 4 + 4 + 5 + 5 + 7 + 8 + 9 = 48. \text{ And then } (48)^2 = 2304.$$

$$\text{Thus the } (\Sigma X)^2 \text{ equals } \boxed{2304}$$

As you might guess, there are numerous other ways that the rules of algebraic order and summation notation can be applied in different statistical formulas. While the calculation of the sum of all squared X values and/or the sum of X squared values are required in many statistical formulas, there are other applications that are also widely used. As such, and to assist in later presentations, we will briefly review a few of the more commonly used applications.

The following two (sub)formulas (or variants of them) are used in several statistical formulas. While both of these formulas instruct us to multiply the X value by the Y value, when this operation is actually completed, it is once again wholly dependent upon the rules of algebraic order. (Unless otherwise specified, when values are found directly next to each other, this mathematically denotes multiplication.)

$$(\Sigma X)(\Sigma Y) \qquad (\Sigma X)^2(\Sigma Y)^2$$

Specifically, the first formula tells us to individually add together the X and Y variables, and then multiply these two values together to obtain the final answer: *parentheses*, then *multiplication*. Conversely, the second formula also tells to individually add together the X and Y variables; but before the resultant answers are multiplied together, each must be individually squared, and then multiplied; *parentheses*, *exponentiation*, and then *multiplication*.

Applied to the "bush-in-the-White House" data sets (Table 2.2), each of these formulas, respectively, is mathematically summarized below. Thus, the sum of X multiplied by the sum of Y equals 3600:

$$(\Sigma X) = 1 + 2 + 3 + 4 + 4 + 5 + 5 + 7 + 8 + 9 = \boxed{48}$$

and then

$$(\Sigma Y) = 2 + 2 + 3 + 5 + 7 + 9 + 11 + 11 + 12 + 13 = \boxed{75}$$

$$(48)\,(75) = \boxed{3600} = (\Sigma X)\,(\Sigma Y)$$

However, the sum of X squared multiplied by the sum of Y squared equals 12,960,000:

$$(\Sigma X) = 1 + 2 + 3 + 4 + 4 + 5 + 5 + 7 + 8 + 9 = \boxed{48}$$

and then

$$(\Sigma Y) = 2 + 2 + 3 + 5 + 7 + 9 + 11 + 11 + 12 + 13 = \boxed{75}$$

$$(48)^2 = \boxed{2304}\ ; (75)^2 = \boxed{5625} \text{ and then}$$

$$(2304)\,(5625) = \boxed{12{,}960{,}000} = (\Sigma X)^2\,(\Sigma Y)^2$$

While all of the statistical formulas found in this text are composed of the above sub-parts (or slight variants of them), to summarize our discussion of algebraic order and summation notation we offer the following formula. This formula is an actual statistical measure of what is called sample variance (applied to a dependent variable [Y] in this example) and is discussed in detail in a subsequent chapter. For now, let's just focus on the order of the mathematical operations called for in this formula.

$$\text{Sample Variance} = \frac{\Sigma Y^2 - \dfrac{(\Sigma Y)^2}{n}}{n - 1}$$

As you will note, this formula has two dividing points in its presentation. As most of you probably know, this tells us to divide the numerator portion of the formula (symbols above the line) by the denominator (symbols below the line). Moreover, these dividing points—lines—also serve as a variant form of parentheses. That is, the operations above and below these lines must be completed before this type of division is done.

To this end, we obviously have to determine three different values represented in this formula; the sum of all squared Y values ΣY^2, the sum of Y values squared $(\Sigma Y)^2$, and the newly introduced letter n that appears twice in this formula. As you will recall, the first two sub-formulas, now applied to a dependent variable (Y), were previously calculated using the "desire" (X) data set. The lower-case letter n represents the size of a given sample data set (versus an upper-case N that represents a population of numbers).

With this in mind, and using the previously offered "usage" data set (Table 2.2), let's go ahead and actually do the calculations required for this formula (sample variance). Working from left to right on the formula, we first determine the sum of all squared Y values; each of the Y values is squared, and then added together.

$$2 \times 2 = 4; \quad 2 \times 2 = 4; \quad 3 \times 3 = 9; \quad 5 \times 5 = 25; \quad 7 \times 7 = 49;$$
$$9 \times 9 = 81; \quad 11 \times 11 = 121; \quad 11 \times 11 = 121; \quad 12 \times 12 = 144; \quad 13 \times 13 = 169;$$

and thus

$$\Sigma Y^2 = 4 + 4 + 9 + 25 + 49 + 81 + 121 + 121 + 144 + 169 = \boxed{727}$$

Next, we calculate the sum of Y values squared; all the Y values are added together, and then squared. This is done below.

$$(\Sigma Y)^2 = (2 + 2 + 3 + 5 + 7 + 9 + 11 + 11 + 12 + 13)^2 = (75)^2 = \boxed{5625}$$

The n or sample size, once again, is the number of research subjects measured. In the present case, there are 10 research subjects/dogs (measurements of Y); thus, $n = 10$. Since we now know each of the values called for in the formula, we simply plug them into it and get the following.

$$\frac{727 - \frac{(75)^2}{10}}{10 - 1} = \frac{727 - \frac{5625}{10}}{9} = \frac{727 - 562.5}{9}$$
$$= \frac{164.5}{9} = \boxed{18.277\overline{7}}$$

Thus, the final answer is $18.277\overline{7}$.

Before proceeding, three brief observations are warranted concerning the above calculations. First, instead of giving confusing rules of when to round up or down when fractions are encountered, in this text we simply carry all calculations to the fourth decimal point and report final answers exactly as they are found on the calculator's register. Thus, we report the above answer to be $18.277\overline{7}$. This brings us to our second noteworthy observation; the bar above the very last value found in this answer, as many of you know, tells us that the 7 repeats infinitely. Finally, and perhaps most importantly, if you can do the above operations correctly, you should take this to be a very promising sign. After all, the rest of the text is simple variations of the above calculations. If, however, any of the above material is confusing, please review it before proceeding.

Box 3

Some statistical formulas require use of a further summation notation procedure. Using the letters *i* and *N*, this additional type of notation instructs us where to begin and end summing numbers from a given data set. More specifically, *i* represents an observation (or set of observations) of a given variable. When a numerical value is placed next to an *i* with an equal sign in between, such as the formula below with i=3, this tells us to begin summing the given set of numbers at the third observation. Conversely, the *N* (as you might guess) indicates which observation to end the summing with. Thus, also found in the formula below, an $N = 7$ instructs to stop summing at the seventh observation. In other words, as the title indicates, the formula below instructs us to sum the *X* of *i*'s from 3 to 7.

Sum of the X of *i*'s from 3 to 7: $\Sigma X\, i_{\,i=3}^{\,N=7}$

To put this into English, we reintroduce the desire/usage data sets (Table 2.3) to assist in this discussion. Additionally, some of the intermediate determinations for the formula above and the one that follows are also found in this table.

To sum the *X* of *i*'s from 3 to 7 in this data set, as graphically represented above, the third through the seventh observations are simply added together. In other words, we simply add 3 + 4 + 4 + 5 + 5 to obtain the final answer of 21.

Alternatively, as the formula below instructs us, let's say that we want to know the sum of the *Y* of *i*'s from 4 to 9 for the "usage" data set.

Sum of the *Y* of *i*'s from 4 to 9: $\Sigma Y\, i_{\,i=4}^{\,N=9}$

As reported above, the sum of fourth though the ninth observations of the above *Y* data set, 5 + 7 + 9 + 11 + 11 + 12, equals 55.

Although this additional summation notation procedure is not found in any further presentations in this textbook, many instructors and other texts assume that you are familiar with it.

LEVELS OF MEASUREMENT

Summation notation and the rules of algebraic order can be applied to any set of numbers. When these procedures are found in a given formula, however, there are often limitations on the types of data sets that can be used. More specifically, some formulas require that certain types of variables be used in their calculation. While one way variables are distinguished in a given formula is to use letters such as *X* and *Y* (representing, respectively, an independent and dependent variable), what are called "levels of measurement" must often also be considered.

As previously noted, social scientists set out to measure different characteristics.

TABLE 2.3 Additional Summation Notation Procedures Applied to Bush-in-White House Data Sets

	Desire a "bush" in the White House (X) Independent Variable			*Number of Times a "bush" was Used (Y) Dependent Variable*
Individual	(X)			(Y)
1	1			2
2	2			2
3	3	3 ← Starting		3
4	4		Point → 4	5
5	4			7
6	5			9
7	5	7 ←		11
8	7	Ending		11
9	8		Point → 9	12
10	9			13

Sum of the X of i's from 3 to 7 = 21

Sum of the Y of i's from 4 to 9 = 55

Characteristics can be measured as constants or variables (this observation and those that follow concerning levels of measurement are summarized in Figure 2.1). An example of a constant in the physical sciences is the temperature at which water freezes: 32 degrees Fahrenheit or 0 degrees Celsius. That is, under normal circumstances, water always—constantly—freezes at this temperature.

In the social sciences, however, there are no real constants. Thus, social scientists are almost exclusively interested in measuring variables. Once again, a variable is any characteristic that changes from element to element in some measurable manner. While the range of possible variables that can be measured in the social sciences is limited only by the given researcher's imagination, the manner in which characteristics are measured always assumes one of three basic forms: nominal, ordinal, and interval-ratio. The abbreviation NOIR is an easy way of remembering these three terms in the order they occur. These types of variables, discussed in turn below, are also referred to as levels of measurement.

Nominal variables, also referred to as *qualitative* variables, are social characteristics that have no real numerical meaning. For instance, things such as hair or eye color, gender, religious orientation, country or state born in, skin color, and so forth, are all variables measured in terms of non-numerical qualities. And although numerical values can be placed on these characteristics so that different groups can be numerically compared, they still have no real mathematical meaning. Thus, while a researcher might be interested in comparing differences between men and women in terms of the number of times they used a bush in the White House or scores on a recent exam, neither gender can have a numerical value placed upon it in any meaningful sense. In sum, the nominal level of measurement simply differentiates people by using non-numerical social characteristics.

FIGURE 2.1 Levels of Measurement

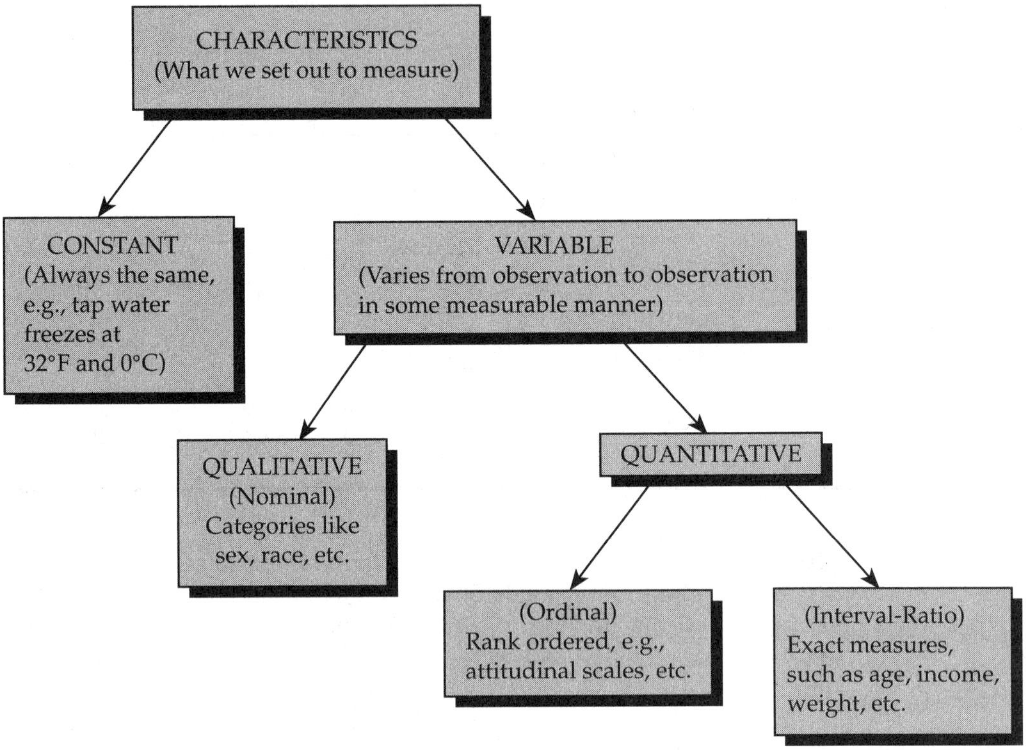

Ordinal and interval-ratio variables not only differentiate social characteristics, they also assign numerical ranks to any measurable differences that are found between observations. In other words, while nominal variables use qualitative forms of measurement, ordinal and interval-ratio variables use quantitative (numerically ranked) forms of measurement. As such, they are also referred to as *quantitative* variables.

Ordinal variables, as the term implies, are concerned with social characteristics that can be *ordered* in some meaningful manner. That is, an ordinal level of measurement numerically scales (rank orders) observations associated with a given social characteristic.

Many social surveys include ordinal questions in terms of respondents' expressed attitudes about a given subject. For instance, we could administer a questionnaire that includes an item that asks subjects to respond to the statement, "All colleges and universities should require students to take a statistics course to graduate." While there are many responses we might receive, questionnaires typically offer a set of standardized responses that are ranked. Standardized responses, also referred to as closed-ended questions, simply mean that possible subjects' responses are predetermined.

One widely used set of standardized answers measures respondents' reported levels of agreement or disagreement on a given topic using a five-point scale, i.e., strongly agree = 1, agree = 2, neutral = 3, strongly disagree = 4, and strongly disagree = 5. Although none of these responses in itself has any numerical meaning, when they are ranked ordered, each value relative to the others does take on a numerical meaning. Thus, we can numerically differentiate between a respondent who strongly agrees (1) with the above statement (probably someone who has already passed the statistics class you are currently enrolled in) versus someone who strongly disagrees (5) with it (most students prior to taking this class).

While ordinal variables measure different rank-ordered responses, they do not tell us the degree of difference between observations. That is, while there is a measurable difference between a respondent who strongly agrees on a given topic versus one who disagrees, the magnitude of this difference is unknown and cannot be determined. Thus, we would have no way of knowing if someone who strongly agrees with the above statement is two, three, or four times more supportive of such a requirement for graduation versus someone who simply disagrees with it.

In order to measure meaningfully the magnitude of difference between various observations we must use what are called *interval-ratio* variables. This level of measurement uses exact measurement intervals to differentiate potential observations. Income, age, weight, and number of times "bush" was used in the White House are all examples of interval-ratio variables. Each one of these variables can be measured using equal measurement intervals: dollars, years, pounds, and number of times used.

Moreover, the measurement units for each of these examples, the intervals, do have meaning in themselves, whereas those associated with ordinal variables are entirely constructed by the given researcher. Thus, it makes sense to conclude that someone who makes $20,000 a year makes exactly twice as much as someone who makes $10,000 a year. Alternatively, the dog who used a "bush" 12 times used it four times as much as the dog who only used it three times. Once again, however, we cannot make these same sort of precise conclusions concerning ordinal or nominal levels of measurement.

Box 4

Statisticians often find it helpful to classify variables also in terms of whether they are discrete or continuous. *Discrete variables* can occur as either qualitative or quantitative characteristics. Mathematically speaking, discrete variables are separate, distinct measurements of a given characteristic. As a result, there are inherent gaps between each observation that cannot be measured. Gender, degree of agreement or disagreement on a given topic, number of children a woman gives birth to, or level of desire for a "bush"-in-the-White House are all discrete measurements. That is, each of these variables is measured in non-overlapping categories that numerically cannot be further broken down.

Thus, while it is common practice to report that women have an average of 2.2 children, it is literally impossible to have 2.2 children. In other words, although a given woman can logically have 2 or 3 children, she cannot have a fraction (.2) of a child.

Continuous variables, on the other hand, can take on literally any value within a predetermined range of numbers. Weight, temperature, or time it takes to use a "bush" are all examples of continuous variables. Each of these variables can be meaningfully broken down into ever finer measurement units. For instance, the amount of time it takes a dog to use a "bush" can be measured in terms of years, days, hours, minutes, seconds, fractions of a second, and so forth. Unlike discrete variables, then, continuous variables can be expressed logically in ever finer measurement units.

In sum, all nominal and ordinal levels of measurement are also discrete variables. An interval-ratio level of measurement, however, can occur as either a discrete or continuous variable. Since social scientists work primarily with discrete variables, our research findings are somewhat limited (see discussion below).

Several summary observations are warranted concerning levels of measurement. This section began by noting that levels of measurement often determine which statistical procedures can be used. The importance of this observation is that some statistical procedures are seen as more powerful than others. Specifically, statistical procedures developed for interval-ratio measurement are seen as most powerful. Next, ordinal statistical procedures are seen as less powerful than interval-ratio procedures but more powerful than nominal procedures. Thus, nominal, ordinal, and interval-ratio variables are respectively seen as having different "levels" of measurement power: from lower to higher.

A related observation is that the three levels of measurement, considered together, constitute an actual ordinal scale. That is, while a nominal variable is seen as a rather inexact way of measuring a given social characteristic, interval-ratio variables are seen as a more precise, definitive form of measurement. Ordinal variables are seen as falling somewhere in between. Accordingly, and at the risk of sounding redundant, these three types variables are referred to as *levels*—rank ordered—of measurement.

Further, since interval-ratio variables are more powerful, for analysis purposes this level of measurement is the most desired. Unfortunately, however, many if not most measurable characteristics available to social scientists are nominal and ordinal. As a result, social scientists regularly and somewhat inappropriately use statistical techniques developed for interval-ratio variables on lower levels of measurement. For instance, it is common practice to analyze an ordinal attitudinal scale as interval-ratio data. While some researchers see no problems in doing this, most correctly recognize that using upper-level statistical procedures on lower-level data increases the likelihood that inaccurate and incorrect conclusions may be drawn.

Finally, there are instances in which social scientists redefine higher levels of measurement into lower ones to make it easier to interpret the results of the given statistical test. For example, years of education, an interval-ratio measure, is often collapsed into an ordinal scale of primary, secondary, undergraduate, and graduate levels of education. Although these four basic categories are not equal, resultant findings from them are sometimes easier to interpret. In most instances, however, because of the already noted increased risk of obtaining invalid results with lower levels of measurement, one should avoid redefining upper levels of measurement whenever possible.

Chapter Summary and Conclusions

This chapter likened statistics to a recipe. More specifically, both recipes and statistics use ordered operations applied to predetermined amounts of ingredients. And if the operations found in a recipe or a formula are not correctly used in the right order, erroneous results will be obtained: burnt food or a wrong answer. An additional discussion of levels of measurement concerning types of ingredients—nominal, ordinal, interval-ratio variables—was also offered in this chapter. Since special attention is given to issues of measurement in many of the chapters that follow, and one must always carefully follow the instructions given with any formula to obtain a correct answer, please carefully review any of the following terms that might be at all unfamiliar to you. This will significantly reduce the possibility of having a "math atheist's nightmare" on the next exam.

Key Terms To Remember

Ingredients
Amounts
Operations
Algebraic Order
Summation Notation
Constants
Qualitative Versus Quantitative Variables
Levels of Measurement—Nominal, Ordinal, Interval-Ratio
Discrete Versus Continuous Variables

Practice Exercises

1. In your own words, what are ingredients, amounts, operations, and order in reference to statistical formulas?
2. The data set (X) found in Table 2.4 represents the number of school terms 10 social science majors have put off taking a required statistics class. The second data set, Y, is

TABLE 2.4 Postponement and Math/Statistics Anxiety

(X) *Number of Terms Taking Statistics Has Been Postponed*	(Y) *Level of Math/Statistics Anxiety*
1	3
2	3
2	3
3	2
3	4
2	4
3	5
3	5
4	5
4	7

each student's measured level of math/statistics anxiety. Use these data sets to complete the following exercises on algebraic order and summation notation.

a. ΣX
b. ΣY
c. $(\Sigma X)^2$
d. $(\Sigma Y)^2$
e. ΣX^2
f. ΣY^2
g. $\Sigma X^2 - 1$
h. $(\Sigma X)^2 - 1$
i. $(\Sigma X)(\Sigma Y)$
j. $\Sigma X i \ {}^{N=7}_{i=3}$
k. $\Sigma Y i \ {}^{N=5}_{i=2}$
l. $\Sigma Y i \ {}^{N=9}_{i=3}$

*Optional questions

3. For each of the following variables, define what level of measurement they are: nominal, ordinal, interval-ratio.

a. age
b. hair color
c. an attitudinal scale
d. height
e. IQ
f. religion
g. your name
h. gender
i. income

4. Now determine if these same variables are discrete or continuous.

Chapter 3

Frequency Distributions, Graphs, and Charts

As noted in Chapter 1, one of the foremost uses of social statistics is to describe social phenomena in terms of variables and the relationships among them. One simple and highly effective way to do this is visually using graphs and charts: summary illustrations of quantitative and qualitative variables. Such pictorial displays are often not only aesthetically pleasing ways to summarize sets of numbers, but they also enable us to visualize things about variables that we might miss by looking at just the numbers. In other words, graphs and charts allow us to view variables and variable relationships in a manner that is often more meaningful and easier to understand. Perhaps this explains their widespread use in magazines, textbooks, and on television.

This chapter explores graphs and charts in terms of how they are constructed and what they represent. Both graphs and charts ultimately represent and are constructed using what are called frequency distributions: predetermined organizational categories that group and summarize observations of a given variable(s). Quantitative versus qualitative variables, however, call for different types of frequency distributions and corresponding graphic presentations, e.g., histograms, frequency polygons, bar charts, and pie charts. As such, we first discuss quantitative and qualitative frequency distributions in terms of what they represent and how they are constructed. Then, these four types of graphic representations are individually discussed.

Quantitative Frequency Distributions

At its most basic level, frequency distributions take raw data, also referred to as *ungrouped* data, and *group* (place) them into predetermined categories. Moreover, for any data set to be grouped in a meaningful manner, it must utilize what are called *mutually exclusive* and *exhaustive* categories. "Mutually exclusive" means that no one observation can be placed into more than one category at a time. Alternatively, "exhaustive" means that there is an available category for each observation. What a frequency distribution does is describe the rates (frequency) of occurrence of variable observations in mutually exclusive and exhaustive categories.

When we are dealing with quantitative variables (ordinal and interval-ratio levels of measurement), the grouping of data is accomplished using numerical categories representing a range of numbers. A quantitative frequency distribution, then, is made up of predetermined numerical categories that are mutually exclusive and exhaustive. Thus, if we had 10 numerical observations of a given variable, to construct a frequency distribution of it, every observation must be grouped into a predetermined numerical category. Moreover, there must be an available category for each observation.

Since what we have said thus far is pretty abstract, let's use the following example to better explain the above terms. Cartoon 3.1 and its accompanying data set (see Table 3.1) represent the number of times an adolescent child has to go "huh" and/or "heh" before its parents become angry. The data set tells us very little other than that there are 25 observations in this sample (n = 25) and the least number of huh/heh's it took to make a parent angry was 2 while the most was 29. Although we have rank ordered the observations from least to most, it is nearly impossible to describe efficiently and summarize 25 separate observations. Constructing a frequency distribution, however, will enable us to arrange this data set into a potentially far more meaningful format. To this end, the first step we must undertake is to determine what sort of numerical categories we will use. These are referred to as class intervals.

Class Intervals

A class interval is a predetermined numerical category that can contain more than one possible observation in its range. Theoretically, the size of a given class interval and its range are arbitrary and literally can be any conceivable values. If, however, we choose an interval that is too large or too small, the resulting frequency distribution will be meaningless.

CARTOON 3.1

TABLE 3.1 "Huh" and/or "Heh" Data Set

Number of "Huh" and/or "Heh" Required to Anger Parent					
2	8	14	17	22	
4	10	14	18	23	
7	12	15	19	24	($n = 25$)
7	12	17	19	28	
8	13	17	20	29	

For example, if we select an interval of 30 covering the range of 1 to 30 for the "huh/heh" data set, all of the observations would fall into one category. Conversely, if we select an interval of 2 covering the ranges of 1 to 2, 3 to 4, and so forth, the resulting frequency distribution would look nearly identical to the actual observations in the raw data set. Either way, the resultant frequency distribution does not organize the data set in a manner that increases our understanding of it.

Most of the presentations of quantitative frequency distributions in the social science literature use an interval size that results in somewhere between 5 to 15 total intervals being used. In the "huh/heh" data set, with its observations ranging from 2 to 29, using an interval size of 5 enables us to place all of the observations into just six categories. While this will be clearly demonstrated in a moment, grouping this data set into six categories (versus 30 observations) will obviously give a very different picture of it.

Where we start the first class interval is also somewhat arbitrary. The beginning interval, however, must always include our first observation(s). Once a starting point is selected, it determines the placement of each subsequent interval. The last interval must always include the final observation(s).

Since the first observation in the "huh/heh" data set is 2, we have set the starting point of the first interval at the numerical value of one. Unless your data set includes a value of zero or a negative value, we suggest that the starting point of the first interval should always be above the value of zero. We also suggest that, whenever possible, the value of one be used as a starting point. This makes it easier to determine the placement of subsequent intervals. Using a starting point of one and a class interval of five, the first grouping (interval) is 1 through 5. Any observations of 1, 2, 3, 4, and 5 are placed into this category. The next interval of five contains the values of 6 through 10, the third interval contains 11 through 15, the fourth contains 16 through 20, the fifth contains 21 through 25, and the sixth and last interval contains 26 through 30. Altogether, the resultant class intervals are found in Table 3.2.

Using six intervals allows us to have a category in which each of our observations will fit. Thus, the intervals used in this frequency distribution are exhaustive. Moreover, since no observation can fit into more than one of these intervals (demonstrated more clearly below), the categories are also mutually exclusive. Next, we must

TABLE 3.2 Class Intervals

Intervals of Five
1 – 5
6 – 10
11 – 15
16 – 20
21 – 25
26 – 30

determine the absolute and relative frequencies of occurrence of observations for each of these intervals.

Absolute and Cumulative Frequencies

We are now ready to start organizing our data into a potentially more meaningful representation. To this end, the first thing we must do is determine the number of observations that occur in each of the class intervals. The rate in which observations occur in any given class interval is referred to as the *absolute frequency*(af). Thus, in the "huh/heh" data set, since there are two observations in the first interval of 1 through 5, a 2 and a 4, its absolute frequency is 2. For the interval of 6 through 10, there are five observations that fit: 7, 7, 8, 8, and 10; thus its absolute frequency is 5. Table 3.3 is a summary of the absolute frequencies for all the class intervals in our distribution.

The *cumulative frequency*(cf) of occurrence refers to the number of observations that have accumulated to a given class interval. In other words, cumulative frequencies are determined by incrementally adding the absolute frequencies together. For instance, the cumulative frequency up to the second interval is determined by simply adding the first and second absolute frequencies together.

TABLE 3.3 Absolute Frequencies

Class Interval	*Absolute Frequency(af)*
1 – 5	2
6 – 10	5
11 – 15	6
16 – 20	7
21 – 25	3
26 – 30	2

TABLE 3.4 Absolute and Cumulative Frequencies

Class Interval	*Absolute Frequency(af)*	*Cumulative Frequency(cf)*
1 – 5	2	2
6 – 10	5	7
11 – 15	6	13
16 – 20	7	20
21 – 25	3	23
26 – 30	2	25

In our frequency distribution, the class interval of 1 through 5 has a cumulative frequency of 2 (the same as its absolute frequency) because we are only interested in observations that occur up to, and including, this first interval. The cumulative frequency for the second class interval is determined, as already noted, by adding the absolute frequencies of the first (1 through 5) and second (6 through 10) intervals; 2 af + 5 af = 7 cf. To calculate the next cumulative frequency, then, the value of 7 is added to the absolute frequency of the next interval (11 through 15); thus, 7 + 6 = 13. The remaining cumulative frequencies are simply determined by adding each subsequent absolute frequency to the previous cumulative frequency: 13 + 7 = 20, 20 + 3 = 23, and 23 + 2 = 25. All of these cumulative values are summarized in Table 3.4; the corresponding absolute frequency values that were used in their calculation are also found here.

Organizing our data set in this format enables us to start to draw some meaningful conclusions about it. For example, just over half of the adolescents (13 out of 25) were able to anger their parents using between 1 and 15 huh/heh's. Alternatively, only 7 of the 25 adolescents were able to anger their parents when 1 to 10 huh/heh's were utilized. Moreover, as one might expect, none of the 25 adolescents required more than 30 huh/heh's to anger his/her parents. (Please note: The cumulative frequency of the last interval must always equal the number of observations in the given data set. If it does not, something has been done wrong and we must go back and recheck our calculations.) In other words, we are starting to get an idea of just how many huh/heh's are required to get most adolescents' parents angry.

Relative and Cumulative Relative Frequencies

Even more can be learned about a given data set by calculating the relative and cumulative relative frequencies. Mathematically, a *relative frequency*(rf) is any absolute frequency divided by the total number of observations. In other words, this type of measurement is concerned with the rate of occurrence found in any given interval relative to all of the occurrences (observations). The easiest way to express this is as a percentage. Thus, for the first class interval of 1 through 5, we take the corresponding absolute frequency value of 2 and divide by the overall number of observations in this data set to get 2/25 = .08 or 8%. The value of .08 or 8% tells us that 8% of all the observations occur in the class interval of 1 through 5. For the next class interval of 6 through 10, the absolute frequency of 5 is divided by 25 to get 5/25 = .20 or 20%. The remaining relative frequencies, summarized with the first two in Table 3.5, are obtained as follows: class interval 11 through 15 gives 6/25 = .24 or 24%; class interval 16 through 20 gives 7/25 = .28 or 28%; class interval 21 through 25 gives us 3/25 = .12 or 12%; and the class interval of 26 through 30 gives us 2/25 = .08 or 8%.

Expressing frequencies of occurrence in relative (percentage) terms allows us to make even more refined observations about our data set. For instance, we can note that for nearly one quarter (24%) of our adolescents it took 11 to 15 huh/heh's to anger their parents. Alternatively, 8% of the adolescents angered their parents with just 1 to 5 huh/heh's.

The final component of our frequency distribution that we need to define and construct is the *cumulative relative frequency*(crf). In a manner similar to our construction of cumulative frequency, the cumulative relative frequency is calculated by taking each relative frequency value and adding it to each preceding cumulative relative value. Thus, as you might expect, the cumulative relative frequency for the class interval of 1 through 5 is simply .08 (8%). The cumulative relative frequency of the next class interval of 6 through 10 is determined by adding .08 + .20 = .28 (28%). Next we take this answer, .28, and add to .24 (for the class interval of 11 through 15) and get .52 (52%); we continue in this fashion until we reach the final class interval. As reflected in Table 3.6, the final cumulative relative frequency should always equal 1.00 or 100%.

(Please note: Sometimes this final answer will be slightly larger or smaller than 1.00. This is due to a rounding error. Assuming that previous calculations have been done correctly, we will rightfully treat it as 1.00. After all, the total cumulative frequencies relative to the number of observations in the data set, 25/25, equals 100% If, however, the total crf is drastically different than the value of 1.00 (e.g., .95 or 1.1), then an error has been made and calculations must be rechecked.)

All of the calculations we have made to this point (af, cf, rf, and crf) are summarized as a complete frequency distribution in Table 3.7. Compared to the original data set, grouping each data set into class intervals of 5 makes it much easier to describe.

TABLE 3.5 Relative Frequencies

Class Interval	*Absolute Frequency(af)*	*Relative Frequency(rf)*
1–5	2	2/25 = .08 (8%)
6–10	5	5/25 = .20 (20%)
11–15	6	6/25 = .24 (24%)
16–20	7	7/25 = .28 (28%)
21–25	3	3/25 = .12 (12%)
26–30	2	2/25 = .08 (8%)

TABLE 3.6 Relative and Cumulative Frequencies

Class Interval	*Relative Frequency(rf)*	*Cumulative Relative Frequency(crf)*
1 – 5	.08 (08%)	.08 (8%)
6 – 10	.20 (20%)	.08 + .20 = .28 (28%)
11 – 15	.24 (24%)	.28 + .24 = .52 (52%)
16 – 20	.28 (28%)	.52 + .28 = .80 (80%)
21 – 25	.12 (12%)	.80 + .12 = .12 (92%)
26 – 30	.08 (08%)	.92 + .08 = 1.00 (100%)

This, in turn, often makes it easier to draw meaningful conclusions about the given data set. For instance, since 80% of the adolescents were able to anger their parents using 1 to 20 huh/heh's, it is apparently quite an efficient strategy for eliciting such a response. Alternatively, 1 to 5 huh/heh's do not appear to work very well; only 2 (8%) of the adolescents were able to anger their parents using this number of huh/heh's. In sum, constructing a frequency distribution is a very simple but effective first step in analyzing any data set.

QUALITATIVE FREQUENCY DISTRIBUTIONS

As already noted, frequency distributions can also be constructed using qualitative variables. For instance, let's say that we are also interested in the types of families our 25 different adolescents come from. Unlike the number of huh/heh's data set, however, family type is a qualitative variable: a nominal level of measurement. As such, numerical values cannot be placed upon observations of this variable in any meaningful manner. This, obviously, eliminates the usage of class intervals as categories to group the data. Thus we simply use groupings of types of families. To demonstrate this more clearly, we offer the data set

TABLE 3.7 Complete Frequency Distribution of Huh/Heh Data Set

Class Interval	*af*	*cf*	*rf*	*crf*
1 – 5	2	2	.08	.08
6 – 10	5	7	.20	.28
11 – 15	6	13	.24	.52
16 – 20	7	20	.28	.80
21 – 25	3	23	.12	.92
26 – 30	2	25	.08	1.00

found in Table 3.8; it represents the rate of occurrence (absolute frequency) for the variable of family type.

Several observations are warranted. First, for each of the family types (categories), the absolute frequency is simply noted. Using these non-numerical groupings, these values reflect the given category's absolute frequency; e.g., seven children come from families where both natural parents are present. Second, the categories we have utilized to group our data set (five family types) are mutually exclusive and exhaustive. That is, there is an available category for every observation to be grouped into, and each observation can fit into only one category.

Finally, although numbered 1 through 5, the order in which these categories is presented is quite arbitrary. For example, instead of using "Both Stepparents" as the first category, any of the family types could be in its place. Moreover, the presentation order of any of the above categories is entirely arbitrary.

Because of this, the above and any subsequent presentation of a qualitative variable is less systematic than that of a quantitative variable and its corresponding class intervals. Mathematically, this limits the depth of any descriptions and conclusions we might make. Nevertheless, we can still calculate the cumulative, relative, and cumulative relative frequencies for this distribution.

Following the same steps as used before, since the absolute frequencies are already given, we first calculate the cumulative frequencies. Once again, the first cumulative frequency is simply the first category's (Both Stepparents) absolute frequency of occurrence: 1. The next cumulative frequency is determined by adding the first category (Both Stepparents) to the second category (One Natural Parent and One Stepparent) to give us 9 + 1 = 10. This process continues (see Table 3.9) until we add the last category and get the expected value of 25 (equal to the total number of observations).

The next task is to determine the relative frequency of occurrence for each category. As before, this is accomplished by taking the absolute frequency of occurrence for each category and dividing it by the overall number of observations in the data set: 25. For the category of "Both Stepparents" we take 1/25 = .04 (4%); this is the rela-

TABLE 3.8 Absolute Frequencies for Five Family Types Data Set

Family Type	*Number of Adolescents*
1. Both Stepparents	1
2. One Natural Parent and One Stepparent	9
3. Father Only	2
4. Mother Only	6
5. Both Natural Parents	7
	25 Total

TABLE 3.9 Absolute and Cumulative Frequencies

Family Type	Absolute Frequency(af)	Cumulative Frequency(cf)
1. Both Stepparents	1	1
2. One Natural Parent and One Stepparent	9	10
3. Father Only	2	12
4. Mother Only	6	18
5. Both Natural Parents	7	25

tive frequency for this category. Then we would take 9/25 = .36 (36%); 2/25 = .08 (8%); 6/25 = .24 (24%); and 7/25 = .28 (28%). All of these operations are summarized in Table 3.10.

The final step is to calculate the cumulative relative frequencies. This is accomplished in Table 3.11 in the same manner as the first data set; each cumulative relative frequency is added to the next relative frequency to get the next category's cumulative relative frequency.

Putting all this information together gives us the complete frequency distribution of family type found in Table 3.12.

Grouping the data in this format once again enables us to make meaningful conclusions about them. For example, the largest category of children comes from reconstituted families; over one-third of the children (36%) come from families where one of their present parents had divorced and remarried. Since the order in which the categories are listed is entirely arbitrary, categories can be simultaneously considered together. For instance, we can easily determine that 96% of the children come from families having at least one natural parent, whereas 40% come from families with at least one stepparent.

TABLE 3.10 Relative Frequencies

Family Type	Absolute Frequency(af)	Relative Frequency(rf)
1. Both Stepparents	1	1/25 = .04 (4%)
2. One Natural Parent and One Stepparent	9	9/25 = .36 (36%)
3. Father Only	2	2/25 = .08 (8%)
4. Mother Only	6	6/25 = .24 (24%)
5. Both Natural Parents	7	7/25 = .28 (28%)

TABLE 3.11 Relative and Cumulative Frequencies

Family Type	*Relative Frequency (rf)*	*Cumulative Relative Frequency(crf)*
Both Stepparents	.04 (4%)	.04 (4%)
One Natural Parent and One Stepparent	.36 (36%)	.04 + .36 = .40 (40%)
Father Only	.08 (8%)	.40 + .08 = .48 (48%)
Mother Only	.24 (24%)	.48 + .24 = .72 (72%)
Both Natural Parents	.28 (28%)	.72 + .28 = 1.00 (100%)

GRAPHING TECHNIQUES

Another meaningful way to present data that builds upon what we have discussed thus far is using pictorial displays in the form of histograms, frequency polygons, bar charts, and pie charts. The old cliché that a picture is worth a thousand words is also true in statistics. (Then again, a cartoon is worth scores of statistics examples!) Each of these techniques, now explored in turn, literally presents a given data set as a picture.

Histograms

One way we can graphically present our data is using a histogram. While all of you have previously seen histograms in magazines, newspapers, textbooks, and on television, you probably did not know that the graphic was called this or how it was constructed. Since histograms can only be constructed using quantitative data, we will simply reuse the "huh/heh" data set and its corresponding frequency distribution in the following discussion.

Histograms are constructed using contiguous vertical bars that represent the absolute frequencies of observations found in predetermined class intervals (cate-

TABLE 3.12 Complete Frequency Distribution of Family Type Data Set

Class Interval	*af*	*cf*	*rf*	*crf*
Both Stepparents	1	1	.04	.04
1 Nat'l/1 Step	9	10	.36	.40
Father Only	2	12	.08	.48
Mother Only	6	18	.24	.72
Both Natural	7	25	.28	1.00

gories). In other words, histograms are simply graphic presentations of the number of observations found in each class interval of a frequency distribution. For these bars to touch and be contiguous, however, we must take mutually exclusive categories and make them connect in some meaningful format. To accomplish this, we need to calculate what are called *upper* and *lower real limits*.

Since our task is to make the interval boundaries touch, we need to identify a common boundary that two intervals can share. The easiest way to do this is to use the point that occurs exactly between two intervals' boundaries. Mathematically, this is determined by taking one class interval's upper limit (top of its range) and adding it to the next interval's lower limit (bottom of its range); this total is then divided by two. The resultant answer is one interval's upper real limit and the other interval's lower real limit.

Applied to the "huh/heh" frequency distribution, the upper real limit for the first interval is determined by taking the upper limit of this class interval and adding to the second class interval's lower limit, and then dividing this value by 2: $(5 + 6 = 11)/2 = 5.5$. The resultant answer of 5.5 is both the first interval's upper real limit and the next class interval's real lower limit. Incrementally repeating this procedure, we add the upper limit of the second interval (10) to the lower limit of the third interval (11), divide this answer (21) by 2, and find the shared boundary of 10.5.

While these and all the remaining calculations are summarized in Table 3.13, there are two values whose calculation warrant further explanation: the smallest lower real limit of .5 and the largest upper real limit of 30.5. In each of these cases, since two class intervals are not listed, a hypothetical one is used. Thus the limits of 31 (for the interval of 31 through 35) and 0 (for the interval of -4 to 0) are used to give us $(0 + 1 = 1)/2 = .5$ and $(30 + 31 = 61)/2 = 30.5$.

We now have all the information needed to construct a histogram. Before doing this, however, there are two more terms we must introduce: the *X* and *Y* axes. The *X axis* (also referred to as the *abscissa*) is the horizontal axis on a graph. With a histogram, the *X* axis is typically used to represent the upper and lower real limit values. Conversely, the *Y axis* (also referred to as the *ordinate*) is the vertical axis; with a his-

TABLE 3.13 Upper and Lower Real Limits

Class Intervals	*Upper and Lower Real Limits*	*Absolute Frequency*
1 – 5	.5 – 5.5	2
6 – 10	5.5 – 10.5	5
11 – 15	10.5 – 15.5	6
16 – 20	15.5 – 20.5	7
21 – 25	20.5 – 25.5	3
26 – 30	25.5 – 30.5	2

togram it is typically used to represent the absolute frequency (af) values. With this in mind and using the previously determined upper and lower real limits, we have constructed a histogram for the "huh/heh" data (see Figure 3.1).

Although the actual construction of our histogram is largely self-explanatory, several important observations are still needed. To begin with, the width of the intervals and the height of the bars for histograms are graphically always represented in equal units. The above intervals, represented using upper and lower real limits, are all in equal units of 5 whereas the given interval's absolute frequency is in units of 1. Thus, the third interval covers the distance of 5 (10.5 to 15.5) and goes up six equal spaces (af = 6). Finally, for aesthetic purposes, since the intersection of the *X* and *Y* axes is portrayed as being 0 (which is often done), the starting point for the first histogram bar on the *X* axis starts just to the right of this intersection.

Frequency Polygons

Frequency polygons also represent class intervals; instead of using bars with two points (upper and lower real limits) to do this, they utilize single points called *midpoints.* More specifically, these single points represent the exact center—middle—of any given interval. Mathematically, we accomplish this by adding together the upper and lower limits of a given interval, and then dividing this answer by 2; (*upper limit + lower limit)/2 = interval midpoint*. Applying this, then, to the "huh/heh" frequency distribution and its first class interval yields (1 + 5 = 6)/2 = 3. Thus, the first interval's midpoint is 3.

FIGURE 3.1 Histogram for Huh/Heh Data

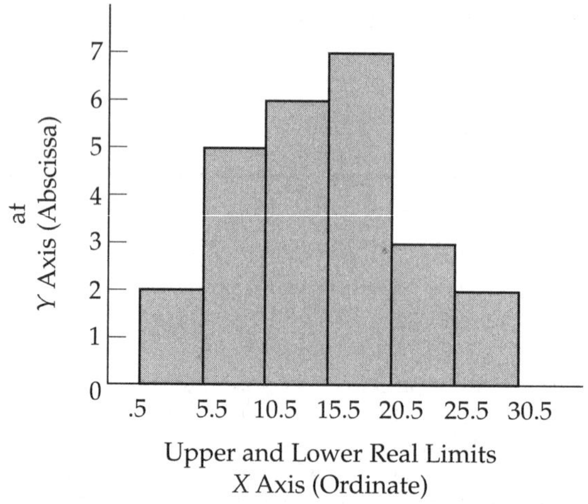

Repeating this procedure with the remaining intervals finds the midpoints of (6 + 10 = 16)/2 = 8; (11 + 15 = 26)/2 = 13; (16 + 20 = 36)/2 = 18; (21 + 25 = 46)/2 = 23; and finally (26 + 30 = 56)/2 = 28. [Please note: The difference between each of these midpoint values (i.e., 3, 8, 13, 18, 23, and 28) is exactly the same as the intervals size: 5. If, incrementally, the differences between all the midpoint values are not exactly the same, the calculations have been done incorrectly. In such instances, obviously, we must go back and correct the error.] All of these midpoint values and the absolute frequencies they represent are summarized in Table 3.14.

The construction of the frequency polygon is very similar to that of a histogram. The absolute frequencies are once again found on the *Y* axis. However, the class intervals on the *X* axis are designated by their midpoints rather than the upper and lower real limits. A simple dot is placed on the graph where the absolute frequency and midpoint intersect. Once all the dots are plotted we simply connect them with straight lines. Finally, at the beginning and ending midpoints the lines are flared to show that the data set does not extend beyond these intervals. Applied to the "huh/heh" data, all of this is found in Figure 3.2.

Bar Charts

We can also construct graphic representations using qualitative instead of quantitative data. One technique that does this with qualitative variables is called a *bar chart*. Although this graphic technique appears to be similar to a histogram, it is plotted very differently. While both represent absolute frequencies on the *Y* axis, since a bar chart represents qualitative characteristics, the *X* axis cannot be labeled in any systematic manner. Thus, the resultant bars represent frequencies of occurrence applied to nonnumerical categories. To demonstrate the construction of a bar chart, we have recreated the family type data (a qualitative variable) in Table 3.15.

Since a bar chart's *X* axis represents qualitative characteristics instead of actual numerical values, each family type must be represented by a single bar. In other words, each family type is represented with a labeled category that also indicates its absolute frequency by the bar's height. Figure 3.3 is a bar chart constructed with the family type data.

TABLE 3.14 Midpoints

Class Intervals	*Midpoints*	*Absolute Frequency*
1–5	3	2
6–10	8	5
11–15	13	6
16–20	18	7
21–25	23	3
26–30	28	2

FIGURE 3.2 Frequency Polygon for Huh/Heh Data

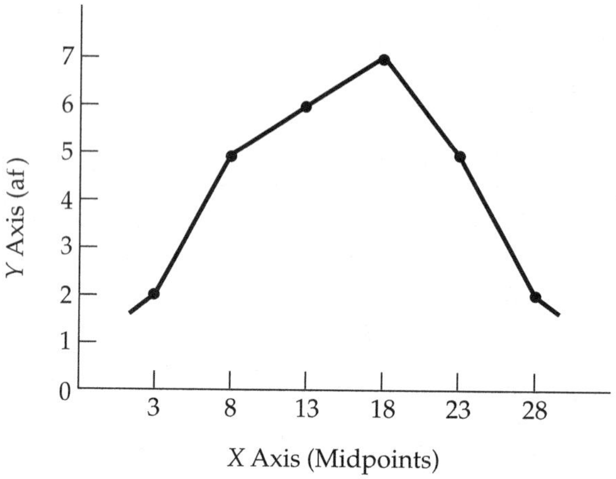

Three things are worth noting about this bar chart and about bar charts in general. First, this technique, unlike that of a histogram, is plotted so that the bars do not touch. That is, unlike ordinal and interval-ratio levels of measurement, since qualitative characteristics (e.g., family type) cannot be scaled, there is no way to display in any meaningful manner categories as touching. This leads to the second point: the order in which the categories appear on a bar chart is largely arbitrary. To be consistent, however, most social scientists will label the bars in the same order they appear in the original data set. Finally, for aesthetic and ethical reasons (see discussion in Box 5), it is preferable for the width of the bars and the distances between them to be equal.

TABLE 3.15 Absolute Frequencies for Five Family Types Data Set

Number of Children for Five Family Types

Family Type	*Number of Children*
1. Both Stepparents	1
2. One Natural Parent and One Stepparent	9
3. Father Only	2
4. Mother Only	6
5. Both Natural Parents	7
	25 Total

FIGURE 3.3 Bar Chart for Family Type Data

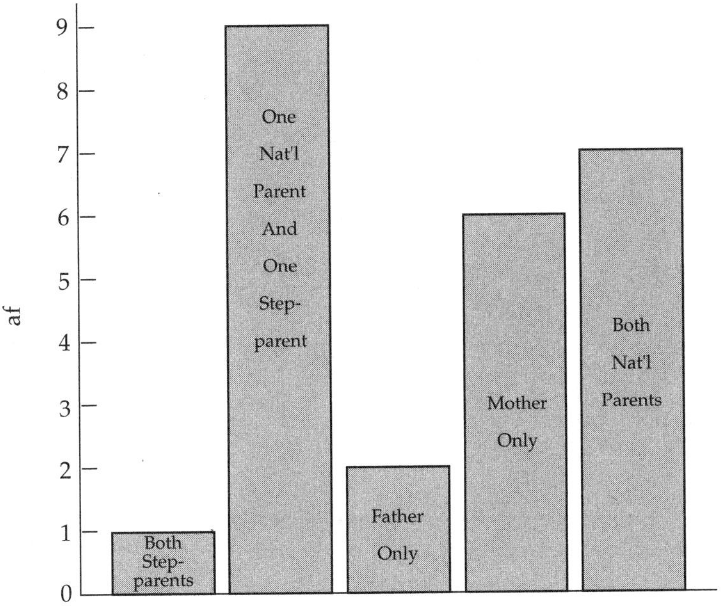

Box 5

While histograms, frequency polygons, and bar charts should all use equal measurement units in their presentation, the actual width and height of the units used is somewhat arbitrary and can give us very different impressions about a given data set. For instance, if one was trying to promote the idea that to anger an adolescent's parents requires numerous, very different amounts of huh/heh's, one would lengthen the distances to portray the absolute frequency and/or shorten the distances used to portray the upper and lower real limits. On the other hand, if one was trying to minimize the differences, one would do the opposite and shorten the absolute frequency distances and/or lengthen the upper and lower real limits that represent the class interval distances. Using a histogram, both of these portrayals are presented in Figures 3.4 and 3.5.

A real-life example of this deceptive ploy is how fluctuations of the unemployment rate are reported in newspapers or on television. If a given reporter or editor is trying to show a drastic increase or decrease in the unemployment rate, the lengths of the bars is increased while the width of the bars is decreased. Alternatively, if he or she is trying to minimize changes in the unemployment rate, just the opposite is done and the lengths of the bars are decreased dramatically while the widths are increased. Pictorially, this is one way that people can lie with statistics.

FIGURE 3.4 Accentuating Differences

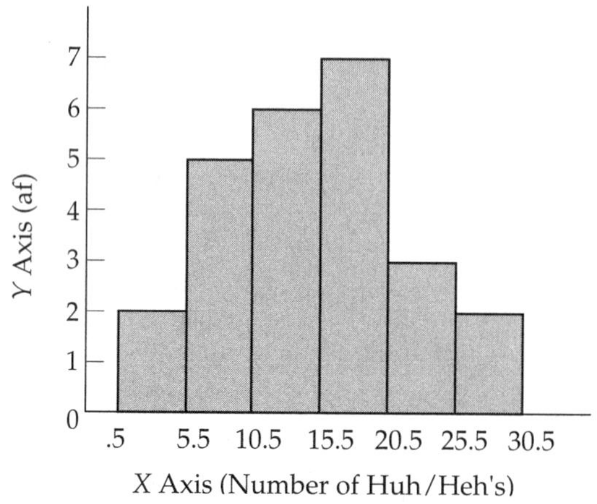

FIGURE 3.5 Minimizing Differences

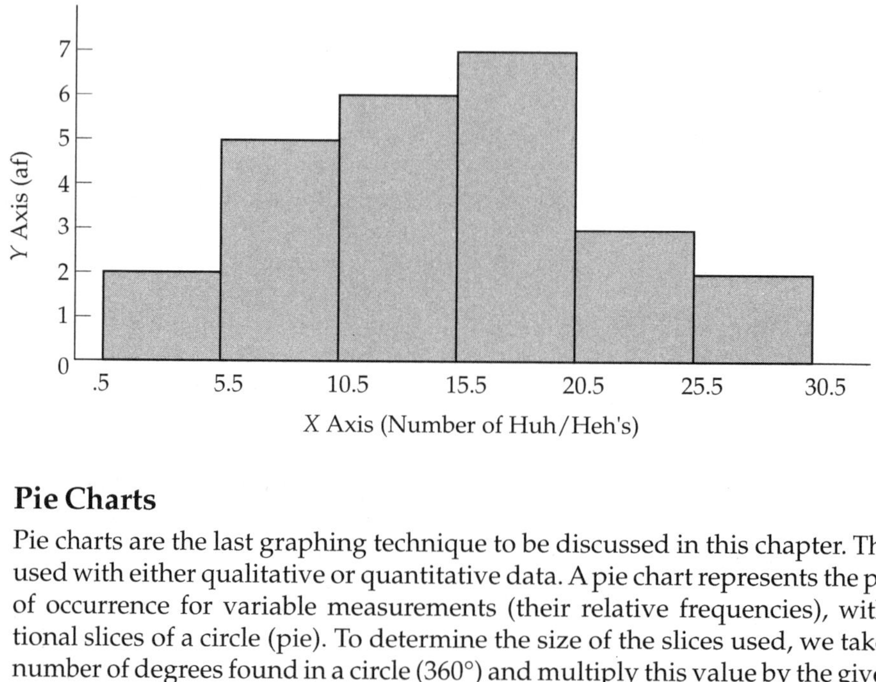

Pie Charts

Pie charts are the last graphing technique to be discussed in this chapter. They can be used with either qualitative or quantitative data. A pie chart represents the percentage of occurrence for variable measurements (their relative frequencies), with proportional slices of a circle (pie). To determine the size of the slices used, we take the total number of degrees found in a circle (360°) and multiply this value by the given category's relative frequency. For instance, if the relative frequency of a category is 25%, we multiply this value times 360 (.25 × 360) and get 90. In other words, a 90° angle is

TABLE 3.16 Slice Size of Class Interval

Class Interval	*Absolute Frequency(af)*	*Relative Frequency(rf)*	*Slice Size in Degrees*
1–5	2	(8%)	$.08 \times 360 = 28.8$
6–10	5	(20%)	$.20 \times 360 = 72.0$
11–15	6	(24%)	$.24 \times 360 = 86.4$
16–20	7	(28%)	$.28 \times 360 = 100.8$
21–25	3	(12%)	$.12 \times 360 = 43.2$
26–30	2	(8%)	$.08 \times 360 = 28.8$
			Total = 360.0 degrees

required to represent 25% of category on a pie chart. All angles are calculated from the very center of the circle.

Replicated in Table 3.16 is the "huh/heh" data with the corresponding class intervals, absolute frequencies, relative frequencies, and calculated degrees of coverage—slice size—required for each interval.

Each category , determined by degrees in an angle, is plotted on an actual circle slice size (see Figure 3.6). Each category is also labeled with both a percentage value and in terms of the class interval it represents. Note that, when all categories have been plotted, the total number of degrees used always equals 360; 100% of all the categories are represented.

FIGURE 3.6 Pie Chart for Huh/Heh Data

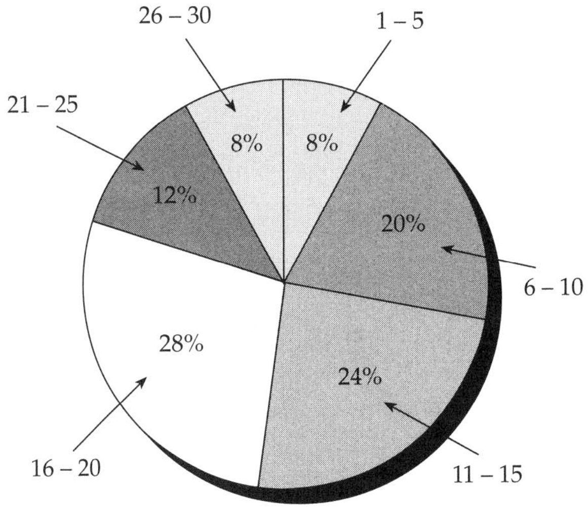

TABLE 3.17 Slice Size of Family Type

Family Type	*Number of Children(af)*	*Relative Frequency(rf)*	*Slice Size in Degrees*
Both Stepparents	1	(4%)	.04 × 360 = 14.4
One Natural and One Stepparent	9	(36%)	.36 × 360 = 129.6
Father Only	2	(8%)	.08 × 360 = 28.8
Mother Only	6	(24%)	.24 × 360 = 86.4
Both Natural Parents	7	(28%)	.28 × 360 = 100.8
			Total = 360.0 degrees

As noted, bar charts can also be constructed with qualitative data. Table 3.17 reports the absolute frequencies, relative frequencies, and calculated slice sizes for the previously used family type data—a qualitative variable.

Following the same procedures used for the "huh/heh" data and using the calculations just made results in the pie chart depicted in Figure 3.7. Instead of numerical intervals as pie labels, however, actual parent types are used.

FIGURE 3.7 Pie Chart for Family Type Data

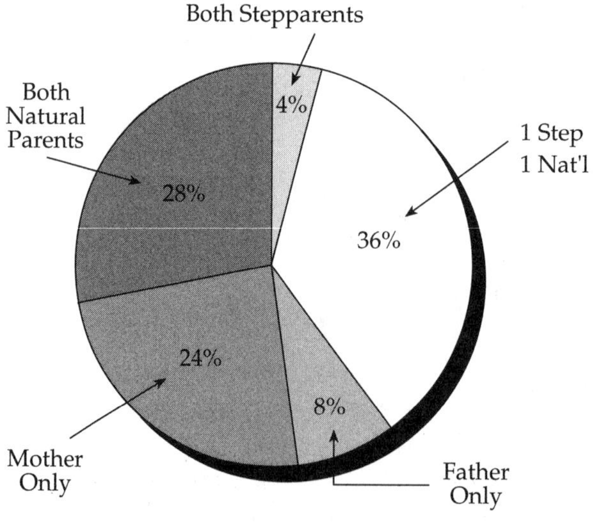

Chapter Summary and Conclusions

Instead of the summary that has been offered at the end of previous chapters, we will review the materials in this chapter using Cartoon 3.2. The homework is also based upon this cartoon. Let's say our current interests are the number of beers that 15 sampled cartoon characters consume in a week's period (such as Hagar, Opus, Zonker, Frank and Ernest, and so forth). The data set in Table 3.18 represents what we found. Note: To assist in subsequent calculations, the data are presented in ascending order from lowest to highest.

CARTOON 3.2

HAGAR THE HORRIBLE reprinted with special permission of King Features Syndicate.

For review purposes, our task is to construct a frequency distribution containing absolute, cumulative, relative, and cumulative relative frequencies using class intervals of three (3). These class intervals with corresponding absolute and cumulative frequencies are found in Table 3.19. (Please note: The intervals start at the suggested value of 1, are mutually exclusive and exhaustive, and when all class intervals are considered together their cumulative frequency equals the sample size.)

A complete frequency distribution also contains the relative and cumulative relative frequencies. Relative frequencies, once again, are calculated by dividing the absolute frequency of each class interval by the sample size (15 in this example). Then, the cumulative frequencies for the intervals are determined by adding each interval's

TABLE 3.18 Cartoon Characters' Weekly Beer Consumption Data Set

	Number of Beers Consumed Weekly by Cartoon Characters		
	1	6	11
	2	7	12
	4	7	12
	4	9	12
($n = 15$)	5	10	15 (Hagar's answer)

TABLE 3.19 Class Intervals and Absolute and Cumulative Frequencies

Class Interval	*Absolute Frequency(af)*	*Cumulative Frequency(cf)*
1–3	2	2
4–6	4	6
7–9	3	9
10–12	5	14
13–15	1	15

relative frequency to the preceding cumulative relative frequency. All of these calculations combined with the absolute and cumulative frequencies are summarized in Table 3.20.

(Please note: Each relative frequency is carried out to the fourth decimal point exactly as found on the calculator. Thus, when the relative frequencies are summed together to determine the last cumulative relative frequency, the final answer equals .9998, instead of the expected value of 1.00. While this is due to rounding errors, observe that when the cumulative frequency of the last category, 15, is divided by the sample size, 15, we get the correct final answer of 100%.)

Now that the frequency distribution is complete, we can construct a histogram and frequency polygon for the data. Remember that to do this we need to calculate upper and lower real limits and midpoints for the class intervals. All of these calculations are summarized in Table 3.21 with the absolute frequencies they represent.

Once again, the upper and lower real limits are calculated by taking one class interval's upper limit and adding it to the next class interval's lower limit, and then dividing this answer by 2 [e.g., $(6 + 7 = 13)/2 = 6.5$]. The lower real limit of the first interval is calculated by taking its lower limit and adding it to the upper limit of the interval that would precede it (0 in this case), and then dividing this answer by 2. With the case in hand, we take $0 + 1 = 1$, divide this answer by 2 to get .5: the lower real limit

TABLE 3.20 Complete Frequency Distribution of Beer Consumption Data Set

Class Interval	*af*	*cf*	*rf*	*crf*
1–3	2	2	.1333	.1333
4–6	4	6	.2666	.3999
7–9	3	9	.2000	.5999
10–12	5	14	.3333	.9332
13–15	1	15	.0666	.9998 (1.00)

TABLE 3.21 Upper and Lower Real Limits/Midpoints

Class Intervals	*Upper and Lower Real Limits*	*Midpoints*	*Absolute Frequency*
1–3	.5– 3.5	2	2
4–6	3.5– 6.5	5	4
7–9	6.5– 9.5	8	3
10–12	9.5–12.5	11	5
13–15	12.5–15.5	14	1

for the first class interval. Conversely, the upper real limit for the final class interval is calculated by adding 15 to 16 (the lower limit of the next class interval that is not presented) and dividing this value by 2: $(15 + 16 = 31)/2 = 15.5$. Midpoints are derived by adding the upper and lower limits of any class interval, and then dividing by 2 [e.g., $(7 + 9 = 16)/2 = 8$].

We now have all the information required for constructing a histogram and frequency polygon for this data set; see Figures 3.8 and 3.9. Remember, for our purposes we have used the *Y* axis to represent absolute frequencies and the *X* axis to represent the upper and lower real limits and midpoints. Also, for our histogram, since the beginning point of our first upper and lower real limit starts at .5, it is placed just to the right of where the *X* and *Y* axes intersect.

To construct a bar chart we need a qualitative data set. As such, we further "asked" our 15 cartoon characters which types of beer they usually drink. Table 3.22 represents a data set of what we found.

FIGURE 3.8 Histogram For Beer Consumption Data

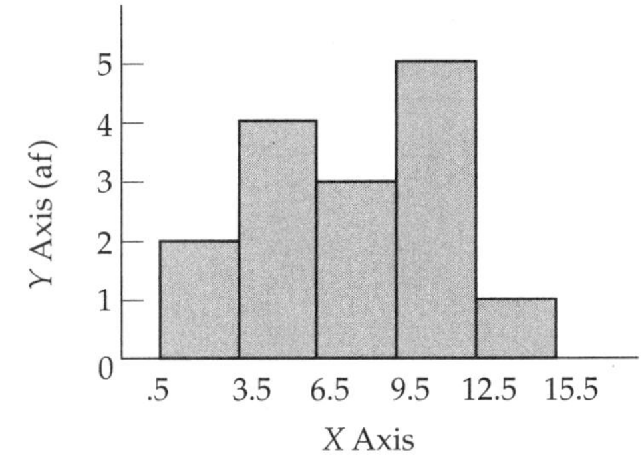

FIGURE 3.9 Frequency Polygon for Beer Consumption Data

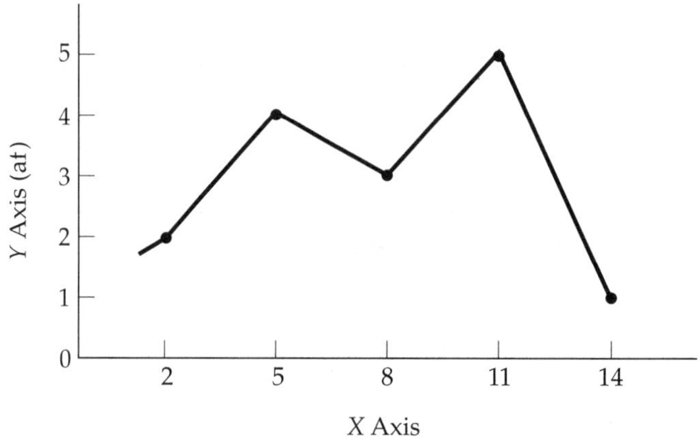

Using this information, we constructed a bar chart: Figure 3.10. (Please note: The bars do not touch and the spaces between them and their widths are equal. The *Y* axis once again represents the absolute frequencies, whereas the *X* axis simply represents the labeled categories in the same order found in Table 3.22.)

As with all of the previous graphics discussed in this chapter, Figure 3.10 gives us a representative picture of what our data looks like. While numbers can speak for themselves, pictures often say things clearer and more forcibly. This is why they can also be used deceptively (see Box 5). Since this ploy is often undertaken by advertisers and the media in general, being conversant in statistics and knowing this, once again, makes it more difficult for others to lie to us.

TABLE 3.22 Brand of Beer Data Set

Brand of Beer	*Number of Characters Who Consume That Brand of Beer*
Bongo Beer	1
Swiller Light	2
Light's-Out Lager	5
Budget Brew	3
Belcher's Pride	1
Cirrhosis Light	3
	15 Total

FIGURE 3.10 Bar Chart for Type of Beer Consumed

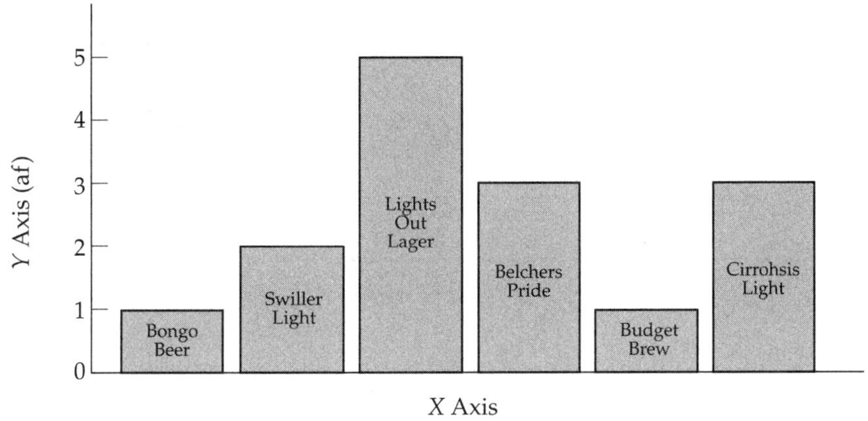

KEY TERMS TO REMEMBER

Quantitative Frequency Distribution	Midpoints
Qualitative Frequency Distribution	Upper & Lower Real Limits
Class Intervals	*X* and *Y* Axes
Absolute Frequencies	Histograms
Cumulative Frequencies	Frequency Polygons
Relative Frequencies	Bar Charts
Cumulative Relative Frequencies	Pie Charts

PRACTICE EXERCISES

1. Using the data found in Table 3.23, construct a frequency distribution using class intervals of 5 minutes; calculate the absolute, cumulative, relative, and cumulative relative frequencies. Present your findings in a table format.
   a. Using the frequency distribution you have just constructed, calculate (1) the upper and lower real limits and (2) the midpoints for each interval.
   b. Using the above information, construct (1) a frequency polygon; (2) a histogram; and (3) a pie chart.

TABLE 3.23 Time Required to Drink Beer Data Set

Number of Minutes Taken to Drink One Can of Beer (First) in the Summertime for 12 Cartoon Characters

2	8	
2	11	
3	12	
6	14	($n = 12$)
7	15	
7	17	

2. Using the data found in Table 3.24, construct a frequency distribution using class intervals of 10 minutes; calculate the absolute, cumulative, relative, and cumulative relative frequencies. Present your findings in a table format.

a. Using the frequency distribution you have just constructed, calculate (1) the upper and lower real limits and (2) the midpoints for each interval.

b. Using the above information, construct (1) a frequency polygon; (2) a histogram; and (3) a pie chart.

3. Let's say we are also interested in the brand of beer consumed by these 12 cartoon characters. Instead of the Midwest (perhaps St. Louis) where the previous data were apparently collected (found in the chapter summary), the above cartoon characters live in the Pacific Northwest. We find the data listed in Table 3.25.

a. Construct a frequency distribution for this qualitative data set.

b. Using the above information, construct (1) a bar chart and (2) a pie chart.

TABLE 3.24 Beer and Visits to the Bathroom Data Set

Number of Minutes Until a Visit to the Bathroom is Required After Drinking the First Can of Beer for Our 12 Cartoon Characters

20	47
22	53
27	54
31	57
38	59
45	62

TABLE 3.25 Pacific Northwest Brand of Beer Data Set

Brand of Beer	*Number of Characters Who Consume That Brand of Beer*
Party Time Pilsner	2
Schacht Pale Ale	4
Marathon Malt Liquor	3
Dud's Suds	2
Pappy's Pale Ale	1
	12 Total

Chapter 4

Measures of Central Tendency

The last chapter demonstrated various ways of grouping raw data into frequency distributions and corresponding graphic representations of them. While these transformations gave us some preliminary ideas about what frequency distributions look like, they also involved several values to make these summaries (e.g., upper and lower real limits and histograms).

In this chapter several techniques are explored that allow us to summarize a whole data set using just one value. Moreover, while there are actually several ways a data set can be summarized with a single measure, this chapter describes four measures of central tendency: mode, median, mean, and weighted mean. As in Chapter 3, levels of measurement will have a direct bearing on which of these techniques we can use. These measures are also presented using grouped data. Before turning to each of these newly introduced terms, a brief introduction concerning measures of central tendency is warranted.

Measures of Central Tendency

When analyzing a given data set it is often quite helpful to identify the most typical score in it. Mathematically, the most typical location within a data set is usually defined in terms of the average score most commonly found in the center of a set of numbers. That is, a *measure of central tendency* identifies what is seen as a middle point of a data set. While there are several ways this determination can be made, each of the following measures is ultimately concerned with describing and summarizing an entire data set of numbers with one single score that is seen as located in or near the center of the distribution.

Mode

The mode is the value that occurs most often or frequently in a data set. A good way to remember this term is to equate the "o" in *mode* with the value that occurs most "often." In terms of mathematical notation, we differentiate between a sample mode (a statistic) and population mode (a parameter), as follows. [Remember that parameters describe a given data set that represents a whole population. Sample statistics, on the other hand, both describe a set of numbers and imply what its true value (the corresponding parameter) is in the population from which it is drawn. Thus, when any of the following measures are considered as statistics, they are also crude estimates of population parameters.]

$$Mo_{\bar{x}} = \text{Sample Mode} \qquad Mo_{\mu} = \text{Population Mode}$$

The mode can be used with all levels of measurement. Using Cartoon 4.1 as a backdrop, let's say we are interested in the nominal characteristic of hair color of students in a statistics class that Calvin is presently enrolled in ($N = 50$). The findings are reported in Table 4.1.

CALVIN AND HOBBES by Bill Watterson CARTOON 4.1

The modal category of this nominal distribution is brown hair. Twenty-two of Calvin's classmates have this hair color, which is more than any other color. Since we are treating the above data set as a population of values, designated by a capital N, the final answer with correct corresponding notation is Mo_{μ} = Brown Hair (22).

Instead of hair color, let's say we are interested in the attitudes of students who have successfully taken this class about taking statistics. From past statistics classes, we randomly sample 25 such students ($n = 25$) and ask them whether they strongly agree, agree, are neutral, disagree, or strongly disagree with the statement, "Taking a statistics class should be a requirement for all students to graduate." Remember, attitudinal scales yield an ordinal level of measurement. Our findings are reported in Table 4.2.

TABLE 4.1 Hair Color of Calvin's Classmates

Color	*Frequency*
Red	3
Brown	22
Black	17
Blond	8
	50 Total

The mode in this case, the answer that occurs most often, with a frequency of 14, is strongly agree or 1. Since our data is a sample of students, the final answer with correct notation is $Mo_{\bar{x}} = 1$ (strongly agree). Since most students strongly disagree with the above statement before taking statistics, successful completion of such a course apparently converts many "math atheists" (see Cartoon 1.1).

Finally, let's say we are also interested in the number of minutes per day our sample of 25 students spent studying statistics to successfully complete this class. Our findings are reported in Table 4.3. With a frequency of 5, the mode for this sample data set is 20 minutes; $Mo_{\bar{x}} = 20$.

Although the mode allows us to report the frequency that occurs most often (in a sense the most popular response), it also has some specific limitations. To begin with, a given data set may not have a mode. For example, let's say the above measured seconds instead of minutes spent studying. This might result in none of the scores being the same. Thus, there would be no mode in this hypothetical example. Moreover, there are also situations where data sets have more than one mode. For instance, in the above if we eliminated the score of 20 minutes, then the frequencies 2 and 3 and their corresponding measurements would be the mode. In situations such as this, the mode is pretty meaningless.

TABLE 4.2 Students' Attitudes About Statistics as a Requirement for Graduation

Attitude	*Frequency*
1. Strongly Agree	14
2. Agree	6
3. Neutral	3
4. Disagree	1
5. Strongly Disagree	1
	25 Total

TABLE 4.3 Absolute Frequencies for Time Spent Studying Statistics

Number of Minutes Spent Studying Statistics	*Absolute Frequency*
20	5
22	2
25	3
30	3
33	4
36	3
37	2
40	2
41	1
	25 Total

Finally, while the mode is the response that occurs most often, it is not necessarily the most representative value of a distribution of scores. The above data set had a mode of 20 minutes, but this single score doesn't tell us much about the rest of the scores in the distribution. The measures of median and mean often give a more accurate representation of a whole distribution of scores.

Median

The median in a data set is similar to a median on a highway. A highway median usually divides a roadway in half. Likewise, a statistical median is the value in a data set which divides a set of observations in half; $\frac{1}{2}$ of the observations are located above the median and $\frac{1}{2}$ are found below it.

Although a median cannot be calculated for a nominal distribution, it can be used for both ordinal and interval-ratio distributions. Although medians are often calculated for ordinal data sets, since such a level of measurement is discrete, answers that result in fractions are somewhat meaningless (e.g., an attitude of 2.5 on a given scale). Thus, continuous data are preferred in calculating medians (and means). If any these terms is confusing, please refer back to Box 4 in Chapter 2 for a more detailed discussion.

To calculate the median, a data set must first be rank-ordered: arranged in ascending or descending order. The median, then, is simply the value located in the exact middle of a ranked distribution. The following notation is used to differentiate between a sample and population median.

$$Mdn_{\bar{x}} = \text{Sample Median} \quad Mdn_{\bar{\mu}} = \text{Population Median}$$

Using Cartoon 4.2, let's say we are interested in the number of city blocks it takes a dog to "catch" a car. We sample 10 dogs and find the following values (expressed in blocks): 8, 2, 7, 4, 3, 2, 0, 5, 4, and 2. To find the median, the first thing we must do is rank-order our data set: 8, 7, 5, 4, 4, 3, 2, 2, 2, and 0. Once the data set is ranked, we use the following formula to determine the exact middle of it. (Please note: Unlike the first three data sets presented in this chapter that were grouped into frequency distributions, this data set is ungrouped, raw data.)

Formula to Determine the Middle of a Ranked Set of Numbers

$$(n + 1)/2 = \text{Middle of Data Set}$$

In the case at hand, the exact center of the data set is $(10 + 1)/2 = 5.5$. This tells us that the median is found at the 5.5 observation. Put another way, counting to 5.5 observations means the median falls between the 5th and 6th rank-ordered observations. The corresponding values of the 5th and 6th observations are 3 and 4. [Please be careful not to confuse rank-ordered observations (e.g., 5th and 6th) with their actual corresponding values (say, 3 and 4). Although a median is determined using ranked observations, it is expressed as an actual data set value.] Thus, the final step is to take these two values, add them together, and divide by 2: $(3 + 4)/2 = 3.5$. The median ($\text{Mdn}_{\bar{x}}$) for this data set is 3.5.

MOTHER GOOSE & GRIMM by Mike Peters CARTOON 4.2

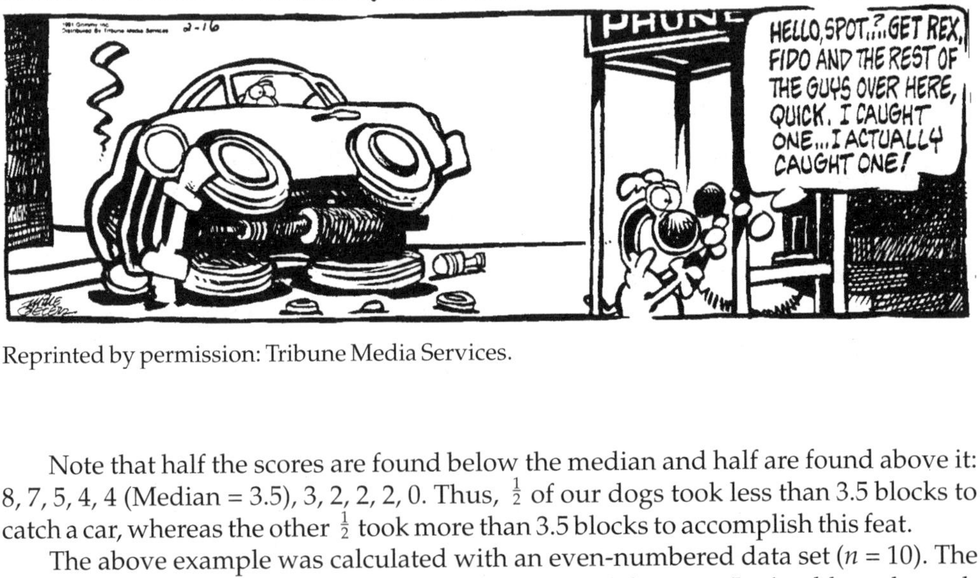

Reprinted by permission: Tribune Media Services.

Note that half the scores are found below the median and half are found above it: 8, 7, 5, 4, 4 (Median = 3.5), 3, 2, 2, 2, 0. Thus, $\frac{1}{2}$ of our dogs took less than 3.5 blocks to catch a car, whereas the other $\frac{1}{2}$ took more than 3.5 blocks to accomplish this feat.

The above example was calculated with an even-numbered data set ($n = 10$). The median is even easier to calculate with odd-numbered data sets. Let's add another value (0) to our data set on dogs chasing cars to make it odd numbered: 8, 7, 5, 4, 4, 3, 2, 2, 2, 0, and 0. Once again using the formula $(n + 1)/2$, we find that the center of this new data set is the sixth observation: $(11 + 1)/2 = 6$. Simply counting to the 6th rank-ordered observation finds ($Mdn_{\bar{x}}$) = 3. In this data set it took the dogs a median of three blocks before they caught the car: 8, 7, 5, 4, 4 (Median = 3), 2, 2, 2, 0, 0.

Mean

The measures discussed thus far only indirectly consider all the values found in a data set: the most typical observation relative to others or the middle of ranked observations. A mean, however, directly considers each value—observation—found in a data set in its calculation. This is accomplished by determining what is called the arithmetic average. Mathematically speaking, then, a *mean* is literally the most "average" value in a data set. As such, it is also often seen as a more exact and representative measure of central tendency. Moreover, in various forms, the mean is one of the most widely used measures in statistics.

To mathematically determine an average, we simply sum all the numerical values found in a data set and then divide this value by the overall number of observations in the data set. Since each numerical value is used in its calculation, the resultant single measure—the mean—literally represents each value. [Conversely, if we take the mean value and multiply it by the sample size ($n\bar{x}$) we find it equals the sum of all the x values.] The formulas to accomplish this for a sample mean ($\bar{x}$) and a population mean (μ) are as follows.

$$\bar{x} = \frac{\Sigma X}{n} \text{ (Sample)} \qquad \mu = \frac{\Sigma X}{N} \text{ (Population)}$$

(called "x bar") (called "mu")

Using the rules of algebraic order and summation notation, both of these formulas tell us to sum all of the X's values and then divide by the total size of the sample or population. A mean cannot be used with nominal data and is predominantly used with interval-ratio data.

Referring back to our car-chasing data (8, 2, 7, 4, 3, 2, 0, 5, 4, and 2), let's say we want to know the mean for this set of observations. We first sum all these values (X of i's), $\Sigma X = 37$. (Note that since a mean adds together all the values, observations do not need to be rank-ordered.) As noted above, the sum of the X's is then divided by the number of observations in the data set; in the present case, 10. Thus, the final answer is $37/10 = 3.7$, or $\bar{x} = 3.7$. Our canines took an average of 3.7 blocks to catch the car.

While the mean is mathematically seen as the most representative value of a data set, there are instances where this conclusion is questionable. In instances where a data set contains an extreme score(s), the resultant mean value may appear to be anything but a point of central location. (Data sets that include extreme values are also often skewed. More will be said about this type of distribution shortly.)

To illustrate, let's say we separately measured 5 additional dogs in terms of how many blocks it took them to catch a car and found the following: 2, 3, 5, 5, and 25. The mean for this data set is 8: $(2 + 3 + 5 + 5 + 25 = 40)/5 = 8$. The mean value of 8 is significantly larger than four of the five observations in this data set. As such, it doesn't appear to be very representative or centrally located although it is mathematically such.

WEIGHTED MEAN

Sometimes we are presented with situations where the mean must take into account another factor in its calculation. For example, let's say we are interested in the average number of seconds it took five different groups of dogs to catch a car. In other words, instead of one data set of dogs, we are interested in the average times of five separately sampled sets of dogs. The findings for our speedy canines are reported in Table 4.4.

We can simply take the average from each group, sum these values and divide their total by the number of groups: (40 + 38 + 35 + 33 + 30)/5 = 35.2 seconds. This is basically how we have calculated means thus far; 35.2 seconds is the simple arithmetic average for all 5 groups. The problem with calculating the average this way is that it does not take into account group size. As such, the value of 35.2 seconds does not accurately represent all five groups.

The easiest way to deal with this problem is to calculate the average by using what is called a weighted mean. A weighted mean compensates for the differing group sizes in its calculation. In other words, it is a more accurate measure in situations where the raw data have been summarized into nominal groups. The notation and formula for the weighted mean are as follows:

$$\text{Weighted Mean} = \bar{x}_w = \frac{\Sigma WX}{\Sigma W}$$

This formula is stated as the sum of the weights times X, divided by the sum of the weights. Many students have trouble deciding what are the weights and what are the X's. Quite simply, since X is the variable, it represents the characteristic of primary concern. On the other hand, W is the a factor that changes X. In the above case, we are concerned with the number of seconds required to catch a car; thus, it is X (e.g., 40, 38, 35, 33, and 30). Since group size potentially changes the overall value of the X's, it then represents the weights (W = 15, 20, 25, 30, and 35). [Recall that the mean of any group multiplied by its corresponding size ($n\bar{x}$) equals the sum of all the X values. Thus, the

TABLE 4.4 Time Taken to Catch Car Data Sets

Group	*Number of Dogs In Group*	*$\bar{X}$ Time (seconds)*
One	15	40
Two	20	38
Three	25	35
Four	30	33
Five	35	30

resultant weighted-mean answer is the same that would be obtained if we used the raw data to make this determination.] To illustrate, let's calculate the weighted mean for the car-chasing data.

$$\bar{x}_w = \frac{(15)(40) + (20)(38) + (25)(35) + (30)(33) + (35)(30)}{15 + 20 + 25 + 30 + 35}$$

$$= \frac{4275}{125} = 34.2 \text{ seconds} \quad \boxed{34.2 \text{ seconds}}$$

By using a weighted mean, we find that the average number of seconds it took the dogs to catch a car for all five groups is actually 34.2 seconds. This is one whole second less than the case in which we did not consider group size.

Box 6

Referring to the actual dog who caught the car in Cartoon 4.2, since she is obviously above average in her capabilities, let's say she hates her day job (chasing cars) and has been attending night school to obtain a bachelor's degree in City Planning. Obtaining such a position would enable her to program traffic lights to make it easier to catch cars. Her grade point averages and the number of credits taken for her four years of attending college are listed in Table 4.5. The school she is attending requires 135 credits with a 2.0 GPA. Will she receive her diploma at the end of this term?

Once again, we can first calculate a simple arithmetic mean by taking the sum of her GPA's and dividing by the number of school years: $(1.50 + 1.75 + 2.00 + 2.50)/4 = 7.75/4 = 1.9375$. A 1.9375 GPA, although close, falls short of the graduation requirement of a 2.0. A more accurate alternative would be to calculate her GPA using a weighted mean which would take into account the number of credit hours earned each year: $[(20 \times 1.50) + (30 \times 1.75) + (40 \times$

TABLE 4.5 Dog's College Grades

Year	*Number of Credits*	*GPA*
First Year	20	1.50
Sophomore	30	1.75
Junior	40	2.00
Senior	45	2.50

(A = 4, B = 3, C = 2, D = 1, F = 0)

2.00) + (45 × 2.50)]/ (20 + 30 + 40 + 45) = 275/135 = 2.037. The correct weighted GPA, enables our fleet-of-foot and oh-so-bright canine to graduate. If your school offers courses for variable credit (e.g., biology 2 credits versus statistics 4 credits) it also calculates your overall GPA using a weighted mean.

Locating the Mean, Median, and Mode in Skewed Distributions

While the level of measurement often decides which measure of centrality we use, the shape of a distribution may also affect this decision. If a distribution is symmetrical, we expect to find the mode, median, and mean all in the same location: equal or nearly equal. Figure 4.1 is a representation of a symmetrical distribution.

Often, however, due to extreme observations in a data set, we are presented with distributions that are skewed. When a distribution is skewed, the mode, median, and mean are found at different locations within it. Distributions that are skewed take two dichotomous shapes: skewed right (positive) and skewed left (negative). Figure 4.2 illustrates each of these distributions.

In both of these distributions the mean is found in what is called the tail region of the distribution. The far left and far right regions of a distribution are considered its tails. Moreover, since both of these figures indicate the presence of extreme scores, the mean is skewed in the direction of these scores.

An example of skewed, positive distribution in the United States is incomes. The average income in this country is always significantly higher than the median income. This indicates that there are a number of individuals (e.g., Donald Trump, Ross Perot) that have extremely large incomes, and as a result they skew the distribution of income in the United States. Perhaps this explains why the median income is most typically used in this country. Instead of having a significant majority of citizens making less

FIGURE 4.1 A Symmetrical Distribution

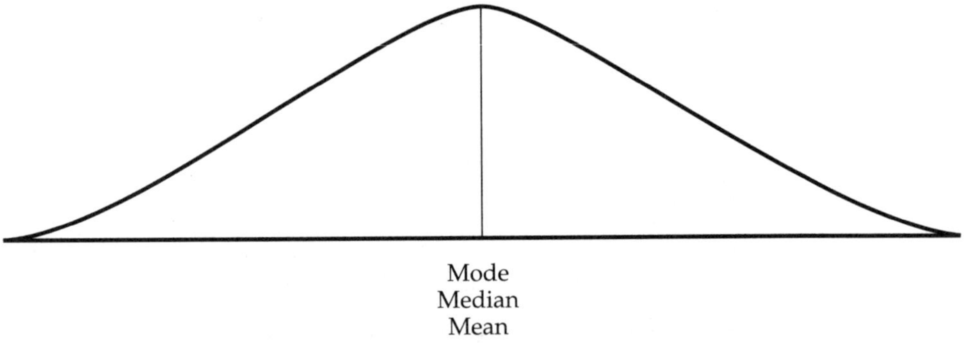

FIGURE 4.2 Skewed Distributions

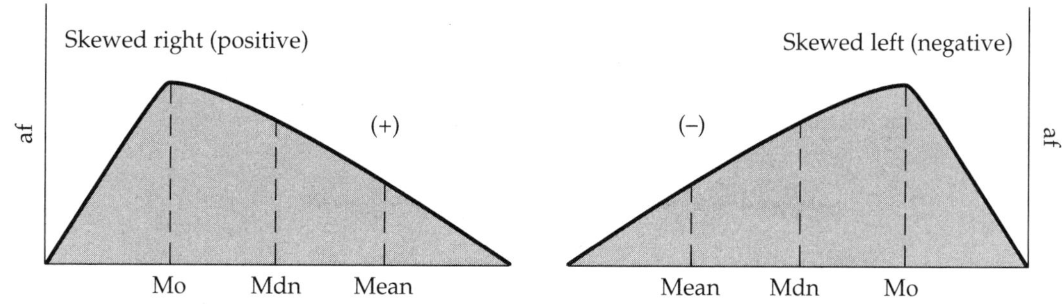

than the average income, only one half of the population ever makes less than the median income. Dependent upon one's ideological outlook, one measure is seen as superior to other.

Box 7

Referring back to Cartoon 4.1, let's say Calvin thinks his weekly allowance of \$2.50 a week is too small in comparison to the amount other kids get. To determine if this is in fact true, we ask the other 15 students in Miss Wormwood's class what their weekly allowances are. Treating this as a population, we obtain the ungrouped, raw data set reported in Table 4.6. Our task is to calculate the mode, median, and mean to determine if Calvin's assertion is in fact correct.

To begin with, the mode weekly allowance is \$5.00. Specifically, four of Calvin's classmates receive this amount—Lisa, Paula, Duane, and Tony—which occurs more often than any other amount. To calculate the median we must first rank-order store prices: 3, 3.5, 4, 4, 4.5, 4.5, 5, 5, 5, 5, 5.5, 6, 6.5, 6.5, 7. Using the formula $(N + 1)/2$ we get $(15 + 1)/2 = 8$. The eighth rank-ordered value has a corresponding value of 5; this is the median. To calculate the mean (μ) we take the $\Sigma X/N = 75/15 = 5$. Considering all three values simultaneously finds:

$$Mo_{\mu} = \$5.00 \quad Mdn_{\mu} = \$5.00 \quad \mu = \$5.00$$

Since all of these values are the same, we apparently have a symmetrical distribution. More importantly, perhaps, from Calvin's point of view we do have evidence that his weekly allowance is much less than other kids in Miss Wormwood's class. Given the amount of money his disruptive behavior probably costs his parents, however, he is probably lucky to receive any allowance!

TABLE 4.6 Weekly Allowances

Weekly Allowances of Calvin's 15 Classmates

Robin	$4.00	Lisa	$5.00	Tony	$5.00	
Sam	$4.50	Paula	$5.00	Brad	$3.00	
Allan	$6.00	Randi	$5.50	Susie	$7.00	($N = 15$)
Harry	$4.00	Jeff	$6.50	Anna	$6.50	
Doris	$4.50	Duane	$5.00	Billy	$3.50	

COMPUTING THE MODE, MEDIAN, AND MEAN USING GROUPED DATA

Sometimes we are presented with data that have been grouped into frequency distributions using class intervals (as discussed in Chapter 3). When data are grouped in this manner and the original raw data are unavailable, we are forced to use variant formulas to calculate the mode, median, and mean.

Referring back to both Cartoons 3.2 and 4.2 as backdrops, let's say we are interested in the average amount of time (in minutes) it takes for a dog to drink—lap up—a beer after chasing a car. Our previous sample of 25 dogs resulted in the findings found in Table 4.7; the raw values used in its construction were unavailable. Our task is to determine the mode, median, and mean using the information found in this frequency distribution.

The simplest way to determine the mode is to find the interval that occurs most often, and then calculate this interval's midpoint. For the above data, the class interval that has the highest frequency, an af of 7, is 16–20; (16 + 20 = 36)/2 = 18, which means the midpoint is 18. Thus 18 minutes is the mode amount of time required to drink a can of beer. Since the raw data are unavailable, however, this is an estimated modal value.

To calculate the median for group data we must first determine the cumulative frequencies for our distribution. This is done in Table 4.8. Remember that when we cal-

TABLE 4.7 Absolute Frequencies for Time Required to Drink a Beer After Chasing a Car

Class Interval	*af*
1–5	2
6–10	5
11–15	6
16–20	7
21–25	3
26–30	1
	n = 24

TABLE 4.8 Absolute and Cumulative Frequencies for Time Required to Drink a Beer After Chasing a Car

Class Interval	*af*	*cf*
1–5	2	2
6–10	5	7
11–15	6	13
16–20	7	20
21–25	3	23
26–30	1	24
	$n = 24$	

culate the median we are looking for the middle observation in a ranked data set. To this end we take $(n + 1)/2$ or $25/2$, which tells us that the 12.5th case is the median. This observation is found in the class interval of 11–15 (cf = 13). However, we do not know exactly where in this interval it would fall. To more accurately make this estimate we use the following formula.

$$\text{Mdn} = \text{Class Interval Lower Real Limit} + \left(\frac{n/2 - \text{cf below}}{\text{af of Mdn Interval}}\right)\text{Interval Size}$$

Applied to our example:

$$\text{Mdn} = 10.5 + \left(\frac{24/2 - 7}{6}\right)5 = 10.5 + (.833\overline{3})(5) = 14.6665$$

Stated in the context of our example, the median time it took our dogs to lap up a can of beer was 14.6665 minutes.

The formula (found below) for calculating the mean using grouped data is similar to that of the weighted mean. However, instead of weights and actual observations (Xi's) we use the absolute frequency for a given interval multiplied by its midpoint. Applied to the beer-drinking example, once again, let's calculate this alternative mean. The intermediate steps for this calculation are reported in Table 4.9.

$$\text{Formula for Mean Using Grouped Data: } \bar{x} = \frac{\Sigma \text{af } m}{n}$$

$$\text{Final Calculations of Mean for Grouped Data: } \frac{347}{24} = 14.458\overline{3}$$

Our 24 dogs took an average of $14.458\overline{3}$ minutes to drink a can of beer after chasing a car. Apparently, car chasing is a strenuous activity that causes many of these dogs to become quite thirsty.

TABLE 4.9 Intermediate Steps for Calculating Mean Using Grouped Data

Class Interval	*af*	*m*	*af m*
1 – 5	2	3	6
6 – 10	5	8	40
11 – 15	6	13	78
16 – 20	7	18	126
21 – 25	3	23	69
26 – 30	1	28	28
	$n = 24$		347

CHAPTER SUMMARY AND CONCLUSIONS

This chapter discussed several measures of central tendency that social scientists use to describe and summarize data distributions. All of these measures locate single points that are seen as most typical in a given distribution. While all of these measures make this determination using different types of information, each has its advantages and disadvantages. Moreover, measures of central tendency are seen as starting points of analyzing data sets. The next analysis procedure usually undertaken is to determine the amount of variability found in a data set, the topic of the next chapter.

KEY TERMS TO REMEMBER

Measures of Central Tendency	Mean
Mode	Weighted Mean
Median	Skew

PRACTICE EXERCISES

1. Let's say we are interested in the age of Calvin's friends. Yes (believe it or not), in addition to Hobbes, Calvin actually does have friends. Calvin's twenty friends' ($N = 20$) ages are reported in Table 4.10, .
   a. What is the *mode, median* (remember to rank-order the data), and *mean* for this data set?
2. Let's say we are interested in the speed, in miles per hour (mph), that our dogs traveled when chasing cars. We sample 15 dogs and find the following (reported in Table 4.11).
   a. What is the *mode, median* (remember to rank-order the data), and *mean* for this data set?

TABLE 4.10 Calvin's Friends' Ages

Age			
12	10	10	9
4	9	11	8
11	8	8	9
6	11	4	8
6	8	6	7

TABLE 4.11 Speed (mph) that Dogs Traveled When Chasing Cars

Speed		
12	17	19
11	10	22
9	26	16
15	14	13
21	7	24

TABLE 4.12 Types of Cars Chased by 15 Dogs

Year, Color, and Make of Car Chased								
Yellow	1969	Ford	Green	1991	Nissan	Turquoise	1993	Toyota
Blue	1991	Chevy	Black	1991	Dodge	Green	1985	Honda
Red	1964	Ford	Blue	1984	Honda	Rust	1952	Chrysler
Purple	1962	BMW	Blue	1991	Ford	Orange	1978	Honda
Green	1983	Honda	Green	1994	Honda	Silver	1979	AMC

3. Listed in Table 4.12 is the color, year, and make of the car the above 15 dogs were chasing.

a. Calculate all possible measures of central tendency—mode, median, and mean—for this data set.

4. Let's say we are interested in the overall average speed of dogs who chase cars in rural, suburban, and inner-city settings. We find the following (reported in Table 4.13).

a. Calculate the weighted and unweighted means for this data set.

b. Which is more accurate?

TABLE 4.13 Average Speeds for Different Groups of Dogs

Group	*Number of Dogs In Group*	*$\overline{X}$ Speed (mph)*
Rural	10	17
Suburban	15	15
Inner City	20	12

5. Let's say we are interested in last semester's grades of the dog who caught the car. We obtain her grades from the registrar's office (reported in Table 4.14).
   a. Calculate the weighted and unweighted means for this data set.
   b. Which is more accurate?
   c. If these were your grades, which GPA would you rather have?

TABLE 4.14 Grades of the Dog Who Caught the Car

	Dog's College Grades Last Semester	
Course	*Number of Credits*	*GPA*
Statistics	4	4
Biology	2	2
Criminology	3	3
English	4	4
Chemistry	3	2
Bowling	1	2
	(A = 4, B = 3, C = 2, D = 1, F = 0)	

Chapter 5

Measures of Variability

In Chapter 4, we discussed different measures of central tendency used by social scientists. While the mode, median, and mean locate important single points within distributions that are most typical and average, they tell us little else. Thus, to get a more complete and accurate picture of a given distribution we need to consider some further summary measures.

This chapter discusses several measures of variability, also referred to as measures of dispersion. To this end, the measures of range, variance, and standard deviation are explored in terms of what they represent and how they are calculated. These three measures are also summarized with single values. A more complete and accurate description of a data set results when both measures of central tendency and variability are used. Moreover, like measures of central tendency, measures of variability are used regularly as sub-parts of numerous statistical techniques discussed in the second half of this text. Before turning to any of these new techniques, however, a brief introduction to variability and why it is an important measure are needed.

Measures of Variability

As the term suggests, measures of variability tell us how much the observations in a data set *vary*, how they are *dispersed* within a distribution. Although measures of central tendency indirectly take into account the amount of variability in their calculation, they cannot assess how a set of observations are dispersed within a distribution. For instance, using Cartoon 5.1 as a backdrop, we are interested in the number of miles a penguin must waddle in the desert before its traveling companion appears to be human. We take eight penguins from a zoo and eight penguins from the South Pole and drop them off in the middle of a desert. Expressed in terms of number of miles it took before its companion appeared to be human, we find the following (reported in Table 5.1). Calculate the mode, median, and mean for each of the data sets.

CARTOON 5.1

Not only are both groups the same size ($N = 8$), but the resultant mode, median, and mean are exactly the same for both groups: 5. If we were using just measures of central tendency to summarize these two data sets, we would conclude that the two distributions are apparently identical. This, however, is obviously not true. That is, while zoo penguins appear inevitably to have hallucinations that their traveling companion is human after 5 miles, there is a great deal of fluctuation (variability) as to when penguins from the South Pole have such illusory images. Thus, to get a more complete picture of a data set, we must know both its measures of central tendency and variability. Moreover, as we will see in later chapters, only when both values are simultaneously considered can meaningful comparisons of different data sets be made.

TABLE 5.1 Number of Miles Traveled before Traveling Companion Appears Human

Zoo Penguins	*South Pole Penguins*
5	2
5	8
5	4
5	7
5	5
5	6
5	5
5	3

TABLE 5.2 Miles Crawled before Traveling Companion Appeared to be a Penguin

Number of Miles			
2	7	13	
4	9	14	
4	10	15	($N = 15$)
4	11	19	
5	13	20	

Range

One of the easiest ways to describe the amount of variation in a data set is to note its smallest and biggest values. Statisticians often refer to these values as X *maximum* and X *minimum*; literally, these are the largest and smallest values found in the data set. The values of X maximum and X minimum are also used to calculate the *range*: a simple measure of the distance between the largest and smallest observations in a data set. Mathematically, this is accomplished by subtracting the smallest value, X minimum, from the largest value, X maximum: *X maximum − X minimum = range.*

To demonstrate this, let's reuse Cartoon 5.1 as a backdrop, and introduce the following as its accompanying data set (Table 5.2). This data set represents a population of 15 humans in terms of the number of miles they crawled across the desert before their traveling companion appeared to be a penguin. The data set is also rank-ordered to make subsequent calculations easier.

With the observations ranked, it is quite easy to see that the largest value, X maximum, is 20, while the smallest value, X minimum, is 2. To calculate the range, X minimum is subtracted from X maximum (20 − 2), to find that this data set has a range of 18. In other words, there is a range of 18 miles in terms of how long it took before the given crawling person's partner appeared to be a penguin.

Often it is helpful to know the highest and lowest scores in a distribution, such as on a recent exam you took. Beyond noting the most extreme scores in a distribution and the distance between them, however, the range has no other value and is not used in any subsequent statistical procedures. As such, it is a very crude measure of the amount of variability in a data set.

Variance and Standard Deviation

While the range tells the width of a data set, it does not take into consideration all of the variability present in a data set. In fact, the only variability it does measure is two extreme values: the largest and smallest. Thus, while the above data set has a range of 18 miles, we have no idea how the other observations are dispersed within it. As a result, we also have no idea if most of the observations occur at the extremes of the distribution (by the 2 and 20) or if they tend to occur towards its center. Two closely related measures that make this determination are called variance and standard deviation. Before we describe how to calculate these measures, however, we need to define more clearly what they represent.

For a measure to mathematically assess the total amount of variability in a data set, it must consider the value of every observation in it. Furthermore, such a measure requires a point from which the observations in the data set can be compared to assess the amount they fluctuate. Since a mathematical mean uses every observation in its calculation and gives us a single point, it is an excellent value to make this determination. That is, one way we can measure the amount of variability in a data set is to use its mean as a point from which each observation is compared.

The easiest way to make this determination mathematically is to subtract the mean from each observation, and then sum these answers into one final measure. The formula that does this, referred to as the *mean deviation*, along with the appropriate calculations applied to the hallucination data set (Table 5.2) are found as follows: $N = 15$ and $\mu = 10$ ($\mu = \Sigma X/N = 150/15 = 10$).

$\Sigma(X - \mu)$ = Mean Deviation Applied to the Hallucination Data Set

$2 - 10 = -8$	$7 - 10 = -3$	$13 - 10 = +3$
$4 - 10 = -6$	$9 - 10 = -1$	$14 - 10 = +4$
$4 - 10 = -6$	$10 - 10 = 0$	$15 - 10 = +5$
$4 - 10 = -6$	$11 - 10 = +1$	$19 - 10 = +9$
$5 - 10 = -5$	$13 - 10 = +3$	$20 - 10 = +10$

Thus,

$$-8 + -6 + -6 + -6 + -5 + -3 + -1 + 0 + 1 + 3 + 3 + 4 + 5 + 9 + 10 = \boxed{0}$$

As most of you have probably noted, the problem with this approach is that the sum of the deviations equals 0. In fact, the sum of the mean deviation for any data set is always zero (0). While this is an interesting outcome, a value of zero is a meaningless measure of dispersion. As such, an alternative method must be employed to measure the amount of variability in a data set.

One simple way to deal with this is to separately square each of the deviation values (e.g., −8 and −6), so that each negative value is canceled out (any negative value squared becomes a positive value). These squared values are then summed and subsequently divided by the number of observations to give us what is considered an average variation value. These operations are all found in the following *definitional formula* of variance.

$$\frac{\Sigma(X - \bar{x})^2}{N} \quad \text{Definitional Formula of Variance}$$

While definitional formulas give us a clearer idea of the operations in the formula, they are also cumbersome to use (especially with larger data sets). As a result, what are called *computational formulas* are usually used. Notwithstanding the above definitional formula that is presented to demonstrate the operations used in calculating the variance (the sum of each observation minus the mean), this text exclusively uses computational formulas. This formula format is much easier to calculate, and although beyond the scope of this text, it is the algebraic equivalent of the definitional formula. (Regardless of which formula is used, the same answer is obtained.)

Listed below are two computational formulas for calculating variance. The first formula, represented by a σ^2 (sigma squared), is for population distributions whereas the second formula, represented with an s^2, is for sample distributions.

Computational Formulas	
Population Variance	*Sample Variance*
$\sigma^2 = \dfrac{\Sigma X^2 - \dfrac{(\Sigma X)^2}{N}}{N}$	$s^2 = \dfrac{\Sigma X^2 - \dfrac{(\Sigma X)^2}{n}}{n - 1}$

The reason for the difference between the formulas (N versus $n - 1$) is that the sample is an estimate. The implications of this are discussed in detail in later chapters on inferential statistics. Nevertheless, since a sample variance always has a smaller denominator ($n - 1$ versus N), it will always result in an answer larger than the equivalent found using the population variance formula. Although it is important to note that these formulas yield different answers, to demonstrate this, and how to make the calculations in general, we will reuse the hallucination data (Table 5.2). That is, although this data set is noted to be a population ($N = 15$), we will also treat it as a hypothetical sample ($n = 15$) for demonstration purposes.

Using the rules of algebraic order and summation notation (please refer back to Chapter 2 if any of the following calculations are confusing), both formulas ask for three initial and identical determinations: (1) What is N or n? (2) What is the sum of all squared X values? (3) What is the sum of the X values? The group size, N or n, of this data set is simply 15. The next determination, the sum of all squared X values, requires that each observation be squared, and then these values are summed together. Alternatively, the last determination, the sum of the X values, requires all the X values to be simply added together. With the column headings of X representing the raw data and X^2 representing each of the observations squared, each of these determinations is summarized below.

X	X^2	X	X^2	X	X^2
2	4	7	49	13	169
4	16	9	81	14	196
4	16	10	100	15	225
4	16	11	121	19	361
5	25	13	169	20	400

$N\ (n) = 15$

Thus,

$$\Sigma X = 2+4+4+4+5+7+9+10+11+13+13+14+15+19+20 = \boxed{150}$$
$$= \Sigma X = \text{Sum of X}$$

and

$$\Sigma X^2 = 4+16+16+16+25+49+81+100+121+169+169+196+225+361$$
$$+400 = \boxed{1948} = \Sigma X^2 = \text{Sum of All Squared } X \text{ Values}$$

Having made these initial determinations, we plug them into their appropriate places on the given formula, and find the following.

Population Variance

$$\sigma^2 = \frac{1948 - (150)^2/15}{15} = \frac{1948 - 1500}{15} = \frac{448}{15} = \boxed{29.866\bar{6}}$$

Sample Variance

$$s^2 = \frac{1948 - (150)^2/15}{15 - 1} = \frac{1948 - 1500}{14} = \frac{448}{14} = \boxed{32.0}$$

As expected, the sample variance using the smaller denominator value (14 versus 15) results in a larger final answer: 32.0 versus $29.866\bar{6}$.

Although the variance is used as a sub-part in many statistical formulas that are explored in the remaining chapters of this text, as a single summary measure it has very little meaning until it is converted into a standardized score called a *standard deviation*. This is mathematically accomplished by simply taking the square root of the obtained variance value. In other words, to calculate a standard deviation value we automatically calculate the variance value. Since a variance involves the squaring of values in its calculation to determine the total amount all observations deviate from the mean, taking the square root of it converts it back into the original units of measurement. The formulas for population and sample standard deviations appear below with the remaining calculations applied to the hallucination data set (Table 5.2).

Population Standard Deviation

$$\sigma = \sqrt{\frac{\Sigma X^2 - (\Sigma X)^2/N}{N}} = \sqrt{\frac{1948 - 1500}{15}} = \sqrt{\frac{448}{15}} = \sqrt{29.8666} = \boxed{5.465}$$

Sample Standard Deviation

$$s = \sqrt{\frac{\Sigma X^2 - (\Sigma X)^2/n}{n - 1}} = \sqrt{\frac{1948 - 1500}{14}} = \sqrt{\frac{448}{14}} = \sqrt{32.0000} = \boxed{5.6568}$$

Several observations are necessary at this point. Once again, the σ and the s tell us whether we are calculating a parameter or a statistic. Remember, upper-case Greek letters are used to signify population parameters while lower-case Latin letters are used to denote statistics. Moreover, since we are taking the square root of a value that is seen as squared, the previous square symbols used to denote variance formulas have been dropped. These symbols (i.e., σ and s), however, are not to be confused with any actual calculations. Rather, like previous formulas (and many to follow), these symbols are summaries of formulas that specify what calculations are undertaken.

Referring back to the data set from which these calculations were made, we now know that these people not only took an average of 10 miles crawling on the desert floor before they had a penguin hallucination but they also had a standard deviation of 5.465 miles. As will become clearer in the next chapter, this value (expressed in miles) will allow us to infer what percentage of desert crawlers are this distance from the mean.

For now, however, variance and standard deviation values still allow us to make one important preliminary conclusion about any data set. Quite simply, the larger variance and standard deviation values, relative to the numerical values of the observations in the data set, the greater the amount of variability that is present in the data set. Conversely, smaller variance and standard deviation values indicate less variability. Thus, the first hallucination data set (Table 5.1) where all the observations equaled 5 has no variability; both variance and standard deviation values equal 0. On the other hand, the above data set, relatively speaking, has a great deal of variability in it and this is reflected in its large variance and standard deviation values.

Additional Example for Range, Variance, and Standard Deviation

Before summarizing this chapter it would be helpful to create one more data set around another cartoon example to review each of the formulas. To this end, we offer Cartoon 5.2 (to be used again for the Practice Exercises) and its corresponding data set found in Table 5.3. (Note: It is not rank-ordered.) The data set represents a sample of 20 children who are six years old in terms of the number of times daily they question their parents' authority (e.g., I don't want to go to bed, stop picking my nose, change my underwear, take a bath, take a math class, etc.).

CALVIN AND HOBBES By Bill Watterson — CARTOON 5.2

Following the same order of presentation used at the beginning of this chapter, we will first calculate the range. Again, this is accomplished by taking X maximum (23) and subtracting X minimum (1) from it: $23-1=22$. Thus, the range for this data set is 22.

Since this is a sample ($n = 20$) we will need to calculate a sample variance and standard deviation. To assist us, the computational formulas needed to do this are replicated below. Remember that these are virtually the same formulas except that the formula for standard deviation involves one more step: calculating the square root of the variance.

Sample Variance	*Sample Standard Deviation*
$s^2 = \dfrac{\Sigma X^2 - (\Sigma X)^2/n}{n-1}$	$s = \sqrt{\dfrac{\Sigma X^2 - (\Sigma X)^2/n}{n-1}}$

TABLE 5.3 Number of Times 20 Sampled Children Question Their Parents' Authority In a Day's Period

Number of Times Authority is Questioned				
4	2	5	7	
11	9	16	6	
1	4	3	14	(n = 20)
21	18	12	8	
7	23	11	10	

Again, as these formulas instruct us, we must calculate the sum of the X's and the sum of all squared X values. We already know that the n or sample size is equal to 20. All of the calculations for these formulas are summarized below; please review them and make sure you understand how each one is made.

X	X^2	X	X^2	X	X^2	X	X^2
4	16	2	4	5	25	7	49
11	121	9	81	16	256	6	36
1	1	4	16	3	9	14	196
21	441	18	324	12	144	8	64
7	49	23	529	11	121	10	100

Thus,

$$\Sigma X = 4 + 11 + 1 + 21 + 7 + 2 + 9 + 4 + 18 + 23 + 5 + 16 + 3 + 12 + 11 + 7 + 6 + 14 + 8 + 10 = \boxed{192} = \Sigma X = \text{Sum of } X$$

$$\Sigma X^2 = 16 + 121 + 1 + 441 + 49 + 4 + 81 + 16 + 324 + 529 + 25 + 256 + 9 + 144 + 121 + 49 + 36 + 196 + 64 + 100 = \boxed{2582} = \Sigma X^2 = \text{Sum of All Squared } X \text{ Values}$$

Sample Variance

$$s^2 = \frac{2582 - (192)^2/20}{20 - 1} = \frac{2582 - 1843.2}{19} = \frac{738.8}{19} = \boxed{38.8842}$$

Sample Standard Deviation

$$s = \sqrt{\frac{2582 - 1843.2}{20 - 1}} = \sqrt{38.8842} = \boxed{6.2357}$$

Our final answers tell us that the number of times these 20 six-year olds questioned their parents' authority has a variance of 38.8842 and corresponding standard deviation of 6.2357. Since the mean for this data set is 9.6 (192/20 = 9.6) and the standard deviation is 6.2357, we can infer that most of the children questioned their parents' authority between 3 and 15 times a day. The last conclusion is based on the idea of a normal distribution and how scores are dispersed within it. This is the topic of the next chapter.

CHAPTER SUMMARY AND CONCLUSIONS

We began this chapter by noting that although measures of central tendency provide us with important information about a data set, they give us no idea about how the observations are dispersed. We then discussed three techniques that measure the variability present in a distribution: the range, the variance, and the standard deviation. The range tells us how much variation is in a data set in terms of two extreme values (the largest minus the smallest), but it does not take into consideration the rest of the values in the data set.

Variances and standard deviations give us a much better measure of how much variability is in a data set by considering all the observations in their calculation. The larger the obtained values are for each of these measures, relative to the numbers being considered, the more variability there is in the data set. Conversely, small variance and standard deviation values indicate that there is very little variability in the data set. Finally, as will become apparent in the following chapters, variances and standard deviations are important sub-parts for numerous other statistical procedures.

KEY TERMS TO REMEMBER

Measures of Variability	Range
X Maximum	Variance
X Minimum	Standard Deviation

PRACTICE EXERCISES

1. Let's say we also collect information on number of authoritarian orders (you will do it because I say so or you will do it or else) our children's parents make daily. Our findings are reported in Table 5.4.
   a. Calculate the range.
   b. Calculate the variance and standard deviation using the appropriate formulas ($n = 20$).
   c. Treating this as a population and using the appropriate formulas, calculate the variance and standard deviation.

TABLE 5.4 Number of Authoritarian Orders Parents Daily Give Their Children

Number of Orders				
6	3	2	3	
12	1	11	7	
4	4	2	14	(n = 20)
11	9	6	7	
7	10	3	9	

2. Let's say that instead of 6-year olds we are now interested in the number of times 13 year olds question parental authority in a day's period. Using 25 research subjects, we find the following reported in Table 5.5.
   a. Calculate the range.
   b. Calculate the variance and standard deviation using the appropriate formulas ($N = 25$)
   c. Treating this as a sample and using the appropriate formulas, calculate the variance and standard deviation.
   d. Are these answers different than those found for the six-year olds? If so, why?

TABLE 5.5 Number of Times 25 Thirteen-Year Olds Questions Their Parents' Authority In a Day's Period

Number of Times Authority is Questioned by Thirteen-Year Olds

8	23	5	7	15	
11	29	16	6	27	
10	14	3	14	18	($N = 25$)
21	18	12	8	9	
17	23	11	10	22	

3. Referring back to Cartoon 5.1, let's say our present research interests are the number of penguins a given person sees once s/he starts to hallucinate. We sample 15 people and find the following, expressed in terms of number of penguins seen during desert crawler's hallucination. (See Table 5.6.)
   a. Calculate the range.
   b. Calculate the variance and standard deviation using the appropriate formulas ($n = 15$).
   c. Treating this as a population and using the appropriate formulas, calculate the variance and standard deviation.

TABLE 5.6 Number of Penguins Seen During Hallucination

Number of Penguins			
6	3	2	
12	1	11	
4	4	2	(n = 15)
11	9	6	
7	10	3	

CHAPTER 6

LOCATING POINTS WITHIN A DISTRIBUTION

Chapters 4 and 5 explored techniques for locating and describing points of centrality and variability in distributions. This chapter takes these ideas a step further and uses them to locate and summarize any given observation relative to all observations in a distribution. Specifically, this chapter explores techniques to calculate what are called percentile scores and standardized scores derived from a normal distribution, also called a Z distribution. Standardized scores are also explored in terms of how they can be used to calculate percentile rankings. The measures discussed in this chapter ultimately allow us to compare and summarize where any given observation is located relative to all the observations in the data set from which it was drawn.

Although we have just introduced several foreign but similar-sounding terms—percentiles, standardized scores, and normal/Z distribution—don't panic. As with previous presentations, in turn, each of these rather simple but difficult-sounding concepts is now explored in detail.

PERCENTILE RANKS AND PERCENTILES

At one time or another, nearly everyone in our society has been assigned a percentile ranking. Often percentile rankings are even unknowingly applied to us. A few examples of *percentile rankings* that may have been applied to you include SAT scores, IQ scores, GRE scores, grades, where you graduated in your high school class, income level, and so forth. In each of these examples, a percentile ranking designates where your score occurs—is located—relative to the rest of the scores in the distribution. Thus, a hypothetical person can obtain a score on the GRE (Graduate Record Exam) in the 98th percentile, but—if still in graduate school—have a personal income that places him or her in the 10th percentile of this latter measure. While most people have an intuitive feeling that being ranked in the upper percentiles is typically good whereas appearing in the bottom percentiles is often bad, we would guess that many people have no idea how this measure is actually calculated.

To demonstrate what has been stated abstractly thus far, we offer Cartoon 6.1 and its accompanying data set (Table 6.1). The data set reflects our current research interests: video game scores for 20 ten-year olds. More specifically, this data set gives video game scores for 20 children on a scale of 0 to 50. A score of 0 means virtually no knowledge of video games, whereas a score of 50 means comprehensive knowledge of such games. Thus, Bob with a score of 50 is seen as having perfect knowledge of video games. Alternatively, Steve's score of 14 means that he has very little knowledge of these type of games.

To begin with, let's say we want to know how Alex's score compares to the other kids. In other words, we want to know Alex's percentile ranking. Found below is the formula used to make this determination. The notation to the left of the equal sign (P_r) simply indicates the formula being used. The two symbols to the right of the equal sign (B and N), however, do call for specific computations. The B represents the number of observations found "below" the score for which we are calculating a percentile ranking. Alternatively, the N should look familiar: It simply represents the total number of observations in the data set. The capital N signifies that all data sets used to calculate percentile ranks are viewed as populations rather than samples. The resultant answer of B/N is then multiplied by 100 to give us an actual percentile ranking.

Formula for Calculating Percentile Rankings

$$P_r = \frac{B}{N} \times 100 = \text{percentile rank of } X_i$$

Before we can calculate Alex's percentile ranking, however, we must first rank-order the raw data in either ascending or descending order. This is the only meaningful way the number of scores below (B) the score of interest can be determined. The order in which a data set is ranked largely depends on whether a high or a low score is

CARTOON 6.1

TABLE 6.1 Video Game Scores for 20 Ten-Year Olds

Video Scores

Janice	46	Anna	38	Allan	15	Pina	32
Harry	49	Randi	28	Alex	43	Katie	17
John	27	Dan	24	Nicole	47	Billy	29
Jenny	34	Jeff	22	Roby	41	Steve	14
Jim	23	Chris	42	Helen	31	Bob	50

seen as more desirable in the given context. Since we jokingly think knowledge of video games is a positive characteristic, the original scores have been ranked from highest to lowest. Finally, to assist in subsequent calculations, we have also numbered the scores. Table 6.2 represents our reformatted data set.

We are now ready to actually calculate Alex's percentile ranking. Applying this data set to the above formula, we first note that since we have 20 observations in it, N equals 20. To determine B, once again, we count the number of rank-ordered scores that occur below the one of interest. For Alex, then, since there are 15 scores that occur below his (6–20), $B = 15$.

Plugging these values into the formula below informs us that Alex's score ranks in the 75th percentile of video game scores. Stated slightly differently, what we are saying is that 75% of the other kids' scores fall below Alex's score. Reflected by his score, Alex is obviously pretty knowledgeable about video games.

$$P_r = \frac{15}{20} = .75 \times 100 = \boxed{\text{75th percentile}}$$

TABLE 6.2 Ranked Video Game Scores

Ranked Scores

1.	Bob	50	11.	Helen	31
2.	Harry	49	12.	Billy	29
3.	Nicole	47	13.	Randi	28
4.	Janice	46	14.	John	27
5.	Alex	43	15.	Dan	24
6.	Chris	42	16.	Jim	23
7.	Roby	41	17.	Jeff	22
8.	Anna	38	18.	Katie	17
9.	Jenny	34	19.	Allan	15
10.	Pina	32	20.	Steve	14

Starting with Harry's score, let's calculate a few more percentile rankings with this data set. Once again, N equals 20, while B in this case equals 18 (the number of scores occurring below Harry's). We plug these values into the formula below to find that Harry's score is in the 90th percentile. With a corresponding score of 49, he is obviously quite knowledgeable about video games.

$$P_r = \frac{18}{20} = .90 \times 100 = \boxed{\text{90th percentile}}$$

Finally, let's say we are interested in Dan's percentile ranking. With 5 scores below Dan's, we divide the B of 5 by the N, 20, to find he is ranked in the 25th percentile. All of this is mathematically summarized below. With a score of 24, Dan would appear fairly knowledgeable of video games. In comparison to other kids, however, Dan's score is relatively low; only 25% of the kids had a score less than his.

$$P_r = \frac{5}{20} = .25 \times 100 = \boxed{\text{25th percentile}}$$

Let's say we are also interested in the exam scores of 10 students who are presently using this textbook. The statistics exam administered to these students was worth 100 possible points. To assist with subsequent calculations, their scores (reported in Table 6.3) are already rank-ordered from highest to lowest.

Summarized below are the calculations for Tony's, Tammy's, and Josh's percentile rankings.

$$P_r = \frac{B}{N} \times 100$$

$$\text{Tony's} = \frac{8}{10} \times 100 = \boxed{\text{80th percentile}}$$

$$\text{Tammy's} = \frac{5}{10} \times 100 = \boxed{\text{50th percentile}}$$

$$\text{Josh's} = \frac{2}{10} \times 100 = \boxed{\text{20th percentile}}$$

TABLE 6.3 Exam Scores for 10 Statistics Students

Exam Scores

1.	Kendra	96	6.	Robin	78
2.	Tony	93	7.	Chris	77
3.	Collin	89	8.	Josh	74
4.	Lisa	86	9.	Sarah	71
5.	Tammy	82	10.	Michael	70

As you can see, Tony is ranked in the 80th percentile, Tammy is ranked in the 50th percentile, and Josh is in the 20th percentile. As with the previous video game example, these students scores of 93, 82, and 74, respectively, have been converted into rankings. While ranking scores is often a helpful way to organize and interpret findings, the original scores and the context from which they are derived should not be forgotten.

For instance, let's say that on the last quiz in this class your score was ranked in the 10th percentile. That is, only 10% of the students had a score lower than yours. On the other hand, however, you had 94% right on this quiz, which earned you a solid A. In this example, we imagine you will take the A and who cares how you compare to the other students in the class. In other words, although percentile rankings can help us make important comparisons, we still need to take into account the original scores and the context from which they were drawn.

STANDARDIZED SCORES AND DISTRIBUTIONS: *Z* SCORES

Another way to rank scores is to use standardized scores in combination with a normal distribution. A normal distribution is the theoretical shape in which we expect frequencies of observations to be distributed in a data set. A normal curve is also referred to the bell-shaped curve, a Z-score distribution, and a Gaussian distribution.

To describe where observations are located in a normal distribution involves the use of two previous measures we have explored: the mean and standard deviation. A mean, once again, is the mathematical average and a measure of centrality for a distribution, whereas the standard deviation measures the average amount of variability present. If either of these terms is unfamiliar to you, please refer back to Chapters 4 and 5.

Using these measures in concert, we can calculate what is called a *Z* score for every actual and hypothetical observation in a data set. In Chapter 9, *Z* scores will also allow us to compare the distributions of two or more data sets to discover if there are differences between them. For now, however, we will explore how *Z* scores allow us to determine where a given observation is located in a data set in terms of its probability of occurrence relative to the whole distribution. The formula for calculating *Z* scores is as follows.

Formula for Calculating *Z* Scores

$$\frac{X - \mu}{\sigma} = Z \text{ Score or } Z \text{ of } X$$

Remember, the μ in the formula represents a population mean, the σ represents a population standard deviation, and the X represents a given observation. Mathematically, the formula instructs us to subtract the population mean (μ) from the observation (X) of interest, and then divide this answer by the data set's population standard deviation. The resultant answer is a *Z score* (also referred to as the *Z* of *X*), the number of standard deviations the score is from the mean.

(Please note: Theoretically, *Z* scores always represent a population of observations. Often, however, since a sample theoretically takes on the shape of a normal distribution in samples over 120, many statistics textbooks will use a sample mean and sample standard deviation to calculate Z scores. Either way, we assume that a rather large data set is being used whenever Z scores are calculated.)

Before we discuss Z scores any further, it would helpful to first practice calculating them. To this end, we offer Cartoon 6.2 and the findings that accompany it; $\mu = 100$ and $\sigma = 15$. These findings represent the IQ's of 4th-grade children who are *not* "willfully stupid" enough to drive their wagon over a cliff—poor Hobbes. (Please note: An average IQ of 100 with a standard deviation of 15 are the same values that researchers use to describe human IQ's.)

Average IQ Score for 4th Grade Children: $\mu = 100$

Standard Deviation for 4th-Grade Children: $\sigma = 15$

With this in mind, what is the Z score for a child with an IQ score of 85? 130? 105? 100? In turn, each of calculations is reported below.

$$\frac{X - \mu}{\sigma} = Z; \qquad \frac{85 - 100}{15} = \boxed{-1}; \qquad \frac{130 - 100}{15} = \boxed{+2}$$

$$\mu = 100, \sigma = 15; \qquad \frac{105 - 100}{15} = \boxed{+.3333}; \qquad \frac{100 - 100}{15} = \boxed{0}$$

One should quickly note that the average of a Z distribution always equals 0 in Z

CARTOON 6.2

scores: (100 − 100)/15 = 0. That is, an observation that equals the mean obviously does not deviate from it. Further, when an X (observation) larger than the mean is considered, we always end up with a positive Z obtained, while an X (observation) smaller than the mean results in a negative Z obtained. In later applications it will be important that we differentiate between negative and positive Z scores, so please make careful note of this in all your calculations. Thus, a child with an IQ of 85 is one whole standard deviation below (−1) the mean, whereas a child with an IQ of 130 is two whole standard deviations above the mean (+2). Alternatively, a child with an IQ of 105 is 1/3 of a standard deviation above the mean (.3333).

In addition to telling us how many standard deviations an observation is from the mean in a data set, Z scores are also used to predict the likelihood of a certain value in a data set being observed. This is accomplished by locating our observation, expressed as a Z score, on a normal distribution—also known as a bell-shaped curve (see Figure 6.1). This figure of a normal distribution, like histograms and frequency polygons, represents a frequency distribution. Unlike most histograms and frequency polygons, however, it is perfectly symmetrical—like a bell curve—in shape. Theoretically, 50% of the observations in a normal distribution are found to the right of the mean (also conceived of as the mode and median) and 50% are found to the left.

The bell-shaped curve represents the frequencies of occurrence in a distribution; the higher the line is in relation to the X axis, the more likely this value is to occur in the data set. Thus, as reflected in Figure 6.1, observations on a normal distribution are more likely to occur around the mean, while the further one gets from the mean the less likely a given observation is to occur.

With this figure in mind, let's reevaluate one of the Z-score determinations we just made. Let's now not only determine how many standard deviations a child with an IQ of 85 is from the mean, but also how many children in general have an *IQ of 85 or less*. (Since the example is based upon humans' IQ scores, one could also think in terms of what is the probability that the next person you will meet—perhaps on a blind date—

FIGURE 6.1 Shape of a Normal Distribution

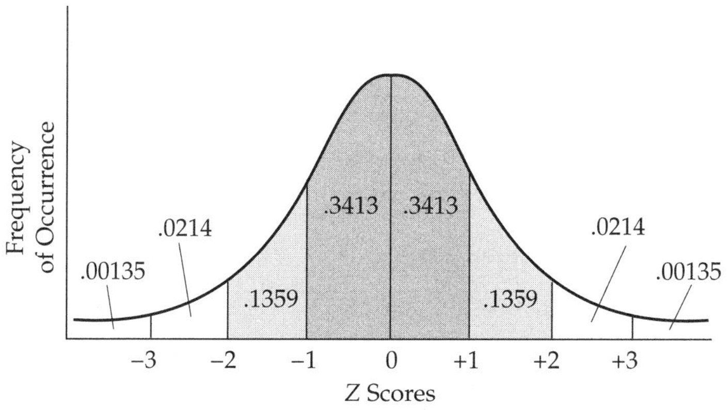

will have an IQ of 85 or less.) When the already determined Z score of −1 for this example [(85−100)/15 =−1] is located on the X axis of the above figure, we find only a small percentage of children with an IQ less than 85. More specifically, a Z score of −1 applied to the above figure tells us that 15.87% of the children have an IQ of 85 or less. We determined this value by noting that the part of the distribution to the left of the mean equals 50% and the distance from the mean to −1 is 34.13%; thus, 50%−34.13% = 15.87%.

Alternatively, we might want to know how many children have an IQ between 100 and 115. Using Figure 6.1, we find that 34.13% of the children have an IQ of 100 to 115. Since 50% of the children have an IQ of 100 or less, we can further conclude that 84.13% of the children have an IQ of 115 or less; 50% + 34.13% = 84.13%.

While many of you may find what has just been abstractly stated to make intuitive sense, from past experience we would guess just as many of you are somewhat lost at this point. Moreover, what do we do when we have Z scores that are not positive or negative 0, 1, 2, 3, etc. (e.g., .3333 or−1.35)? A more systematic and simpler way to make all of these determinations is to use Appendix 1, which contains a summary table of Z scores applied to the normal distribution. Each listed Z value in this table (found in column 1) also offers three corresponding probability values (columns 2–4). We can also use the probability values listed in this table to determine corresponding actual values (*X*).

To demonstrate more clearly this table's usage, we offer the following mean and standard deviation for women's heights in the United States.

Average for Women's Height in the US: $\mu = 64$ inches

Standard Deviation for Women's Height in the US : $\sigma = 2.4$ inches

Let's say we are interested in the percentage of women who are 70 inches or taller. Although there are several different ways of figuring this out, we suggest that you use the following four steps. We have found that students who consistently follow these simple steps significantly reduce the number of potential errors they make.

Step 1: Take the score of interest, *X* = 70, and plug it into the Z-score formula. The resultant answer is also called the *Z of X*. As reported below, a woman who is at least 70 inches tall has a corresponding Z score of 2.5.

$$\frac{70 - 64}{2.4} = \boxed{2.5} = Z \text{ of } X \text{ or } Z \text{ Score for 70 inches}$$

Step 2: Make a graphic representation of where the Z score of interest appears on the normal distribution. This, as we will see, decides which of the four columns in Appendix 1 are used. These columns represent different portions of the normal distribution. Since the mean of a distribution always equals zero and is the distribution's center, all scores to the right of the mean are larger than the mean value and all corresponding Z scores are positive. Alternatively, all observations on the left hand side of the distribution are smaller than the mean and all corresponding Z scores are negative.

In our present example, the value of 70 inches is larger than the mean of 64 inches. Thus, this value in our pictorial representation is to the right of the mean. A large Z score like this one (2.5) tells us that this observation is located well into the far right tail region of the distribution. (Recall that the tail regions are the areas to the far right or far left of the distribution.) Figure 6.2 is an illustration of what this looks like graphically. Since we are concerned with women who are 70 inches or taller, the region of the distribution to the right of our score, $X = 70$ or $Z = 2.5$, has been shaded in.

Step 3: Now that we have a visual idea of where our Z score is positioned on the normal distribution, we locate it in Appendix 1. As previously noted, the first column on each page of the table contains Z scores. Since the Z scores are ranked in ascending order of 0 to 3.70, our Z score of 2.5 is located toward the end of the table. [Please note: Z scores in this table are only carried to two decimal points. Thus, instead of rounding up or down with Z scores that are carried to numerous decimal points (e.g., 1.9578), use the Z score exactly as it appears on the calculator to two decimal points (e.g., 1.95). Using Z scores in this manner still results in quite accurate final answers and saves you from memorizing additional rounding rules.]

Step 4: The final step is to determine which of the three remaining columns in Appendix 1 contains the correct probability value we are interested in, given a Z value of 2.5. This is where our graphic is helpful; it tells us that we are concerned with the tail region of the distribution. Since this region is less than 50% of the distribution, we will simply refer to it as "The Small Part."

In Appendix 1, column 2 is entitled "Z to the Mean." To use this column in the present example, we would have to be interested in the area from the mean, 64 inches, to our score of interest, 70 inches. Obviously, we are not interested in this column. Column 3 is entitled "The Big Part." This column represents areas of the distribution that are greater than 50% of it. In other words, if we were interested in the percentage of women who are 70 inches or shorter we would use this column. As a result, we are also not interested in this score.

FIGURE 6.2 70 inches or Taller

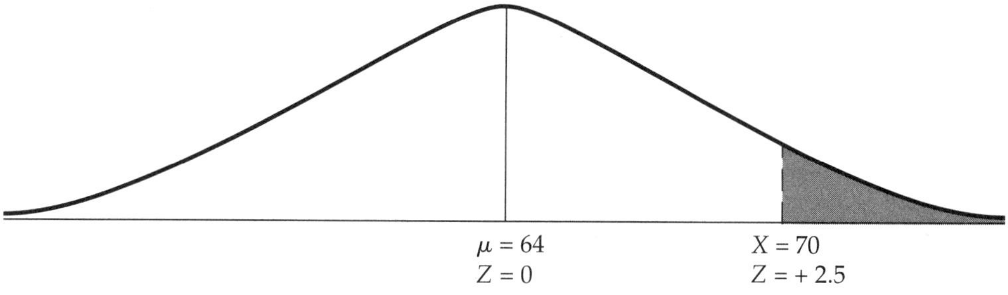

Finally, we have column 4, entitled "The Small Part." This column represents the tail regions of the distribution, the shaded portion of Figure 6.2.

Thus the answer for Step 4, the final answer, found in column 4 is .0062. This tells us that of every 10,000 women in the United States, only 62 are 70 inches or taller. This can also be expressed as a probability; there is a .62% (less than 1%) chance that the next woman you randomly meet (perhaps on a blind date) will be five foot ten or taller. In other words, there are not very many women in the United States who are 70 inches or taller.

To help you in the following examples, the four steps just used are summarized below. (We should note that these are general steps which somewhat more complicated examples will build upon.)

Step 1: Calculate the Z score.

Step 2: Make a graphic representation of the Z of X in terms of what area of the normal curve you are interested in.

Step 3: Locate your Z score in Appendix 1.

Step 4: Determine the appropriate column (columns 2–4) and report the final answer.

Turning to some further examples, let's say we are now interested in how many women in the US are 5 feet tall (60 inches) or shorter. Once again, we first calculate a Z score (Step 1), which is done below.

$$\frac{60 - 64}{2.4} = \boxed{-1.666\bar{6}} = Z \text{ of } X \text{ or } Z \text{ Score for 60 inches}$$

Step 2 instructs us to make a graphic representation of the Z of X. That is, we must draw the normal curve and locate the part of the distribution that we are interested in. Since 60 inches is less than the mean score of 64 inches and the corresponding Z score of −1.6666 is a negative Z value, we are obviously interested in the area to the left of the mean. Further, since we are concerned with the area of 60 inches or less, we shade in the tail region to the left of this score. Both of these steps are presented in Figure 6.3.

Step 3 calls for us to locate the Z score in Appendix 1, which is found in column 1. Recognizing that a Z score is negative is helpful in graphically locating its position on a distribution and will be required in future calculations. At this point, however, when utilizing Appendix 1 we disregard the sign (+ or −) and treat the Z score as an absolute value. In the present example, −1.6666 appears in the appendix as simply 1.66 (recall that we take the Z score exactly as it is found on the calculator register to two decimal points). Like the previous example, since we are interested in a tail region of the distribution, the final answer (Step 4) is located in "The Small Part," column 4. Thus, the final answer is .0485. This enables us to conclude that 4.85% of women in the US are 60 inches or shorter. We can also conclude that the probability that the next woman you will randomly meet will be five feet or shorter is 4.85%, a relatively rare outcome. Moreover, from the past two examples we can also conclude that almost all women in the US are between 60 and 70 inches in height.

FIGURE 6.3 60 inches or Shorter

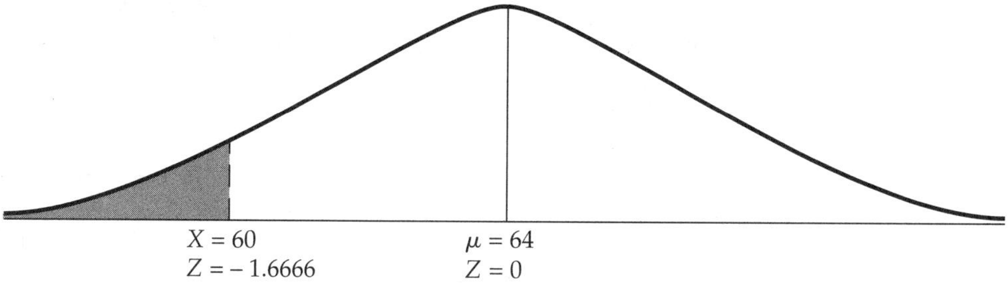

The next example will allow us to utilize one of the other columns. To this end, let's say we want to know what percentage of women in the US are 66 inches or shorter. First, we calculate a Z score.

$$\frac{66 - 64}{2.4} = \boxed{.833\bar{3}} = Z \text{ of } X \text{ or } Z \text{ Score for 66 inches}$$

Our X of interest is larger than the mean and a positive Z score. Thus, we know it is located on the right-hand side of the distribution. Since we are interested the region of the distribution below this point (or shorter), we shade in the area below it. All of these determinations (Step 2) appear in Figure 6.4.

One should immediately note that the shaded-in region covers over 50% of the distribution. Thus, we will be utilizing column 3 in Appendix 1, "The Big Part." Before doing this we must complete Step 3 and locate the Z score of .8333 (actually .83) in column 1. Having done this, we find the final answer (found in column 3) is .7967. This answer tells us that 79.67% of all women in the US are 66 inches or shorter.

For our next example, let's say we are now interested in the percentage of women 61 inches and taller. First we calculate the Z score, and find that the Z of X is a −1.25. A

FIGURE 6.4 66 inches or Shorter

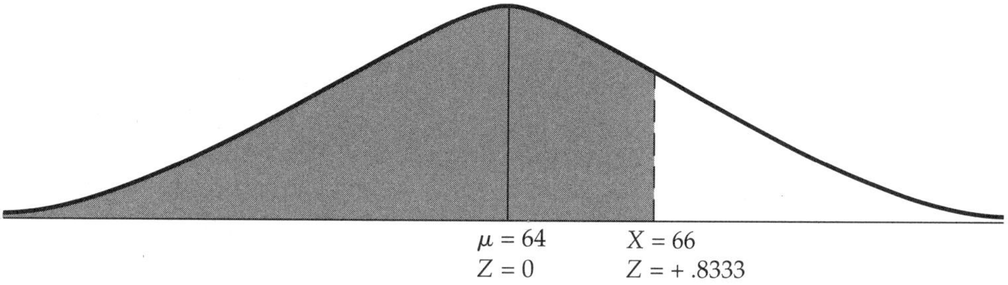

negative score combined with a score that is less than the mean (61 < 64) tells us that the graphic representation of this location point is to the left of the mean. Since we are concerned with the region to its right (61 inches and taller), we shade it in. Both this Z-score calculation and its graphic representation (Figure 6.5) are as follows.

$$\frac{61 - 64}{2.4} = \boxed{-1.25} = \text{Z of X or Z Score for 61 inches}$$

Having completed Steps 1 and 2, we now locate the Z score of -1.25 (1.25) in Appendix 1. Once this is accomplished we find that the final answer (Step 4) in column 3 is .8944. That is, 89.44% of all women in the US are 61 inches or taller.

So far we have used only columns 3 and 4 in Appendix 1. Let's say that now, however, we want to know what percentage of women in the US are between 62 and 64 inches (64 inches is the mean) in height. Again, we start by calculating a Z score, just for 62 inches in this case, and then we make a graphic representation of it (see Figure 6.6).

$$\frac{62 - 64}{2.4} = \boxed{-.8333} = \text{Z of X or Z Score for 62 inches}$$

Instead of shading a complete area of the distribution below or above a point of interest, in this example we shade in the region from our point of interest of 62 inches to the mean: 64 inches. This also tells us that the answer to Step 4 is found in column 2, "The Mean to Z." First, however, we must locate this Z score in column 1 of Appendix 1. Having done this (using the absolute Z score of .83), we find that the final answer in column 2 is .2967. Thus, we conclude that 29.67% of women in the US are 62 to 64 inches in height.

Here's a similar example with a twist. Let's say we are now interested in the percentage of women who are 63 inches to 67 inches in height. This problem is the same as the above except that it is concerned with an area of Z to the mean from both sides of the mean. Thus, we calculate two separate Z scores (Step 1) and then sum their two corresponding probability values. These two Z-score calculations are as follows. The graphic representation of them appears in Figure 6.7.

FIGURE 6.5 61 inches or Taller

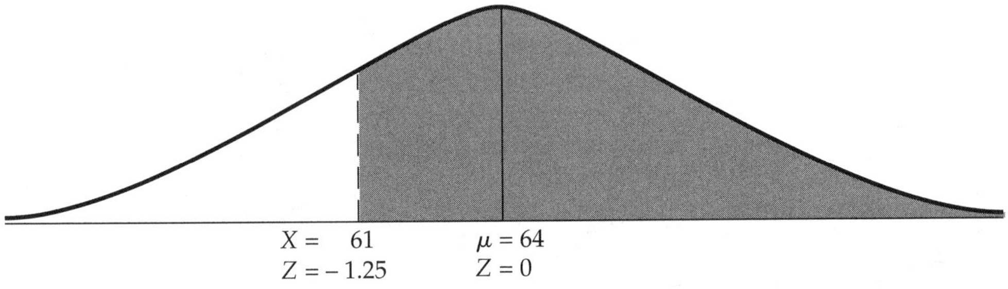

FIGURE 6.6 62 to 64 inches (The Mean)

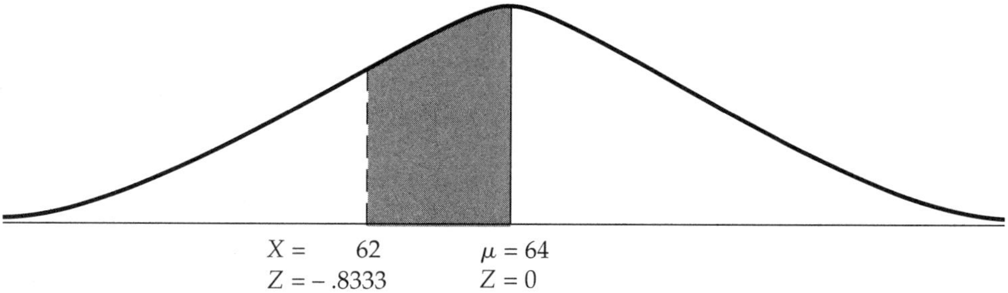

$$\frac{63 - 64}{2.4} = \boxed{-.416\bar{6}} = \text{Z of X or Z Score for 63 inches}$$

$$\frac{67 - 64}{2.4} = \boxed{+1.25} = \text{Z of X or Z Score for 67 inches}$$

We next locate both of these Z scores in Appendix 1: 0.41 and 1.25. Their corresponding probability values in column 2 (The Mean to Z) are then summed together to get .3944 + .1591 = .5535. (In a sense, with the extra operation of summing these two value together, we have added a Step 5.) From this result we conclude that roughly 55.35% of women in the US are between 63 and 67 inches.

Building upon this example, let's say we are now interested in the percentage of women whose heights are between 65 and 70 inches. Here again we are interested in an area between two points on the normal distribution; however, this area falls on just the right-hand side of the distribution. Fortunately, the procedure to make this determination is very similar to the previous example.

Thus, as above, we first calculate two separate Z scores and then locate them on a graphic representation (Figure 6.8). Note that on this representation a sizeable portion of the tail region on the right-hand side of the distribution is appropriately shaded in.

FIGURE 6.7 63 to 67 inches

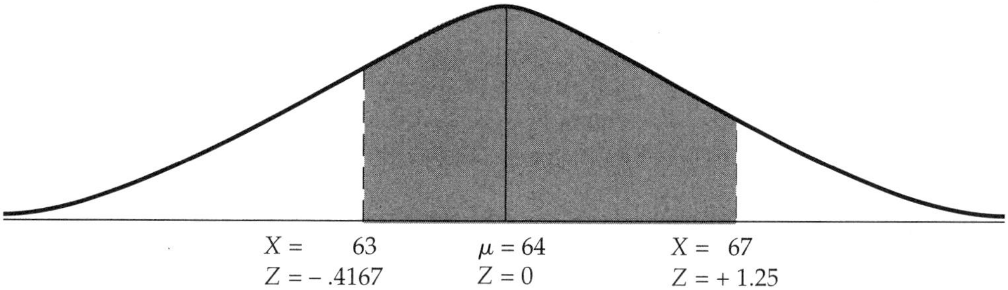

$$\frac{65 - 64}{2.4} = \boxed{+.416\bar{6}} = Z \text{ of } X \text{ or } Z \text{ Score for 65 inches}$$

$$\frac{70 - 64}{2.4} = \boxed{+2.50} = Z \text{ of } X \text{ or } Z \text{ Score for 70 inches}$$

Next, we determine the probability values for both of these Z scores in terms of their distance from the mean (column 2). The corresponding probability values for the Z scores of .41 (65 inches) and 2.5 (70 inches), respectively, are .1591 and .4938. Instead of adding these values together, as above, we subtract the smaller value from the larger (to do otherwise would result in a negative probability value, which is logically impossible). In other words, we are subtracting the probability value of the unshaded area, 64 to 65 inches, from the probability area of the mean to the second score, 64 to 70 inches. Thus, we find that 33.47% of women in the US are between 65 and 70 inches in height: .4938−.1591 = .3347.

(Please note: Sometimes our students make the *mistake* of subtracting 65 inches from 70 or the Z score of .41 from 2.5 to make this determination. Since the area under the normal curve, expressed as a probability value, is unequal, actual probability values *must always* be used to make this determination. Otherwise, as you might expect, an incorrect answer will be obtained.)

Thus far we have been concerned with using Z scores to discover the percentage of the distribution above or below a given X (observation) or between two X's. Sometimes, however, it is useful to do this in the opposite direction and discover what the X value is for a given probability score. The formula to accomplish this, an algebraic derivative of the original Z-score formula, appears below.

$$X = Z\sigma + \mu$$

To use this new formula we basically use the steps for the original Z formula in reverse. Moreover, to make its usage easier, we will always treat the observations in the data set as rank-ordered. That is, the percentages to the left of the mean are seen as less than 50% while those to the right are seen as greater than 50%. Thus, the furthest

FIGURE 6.8 65 to 70 inches

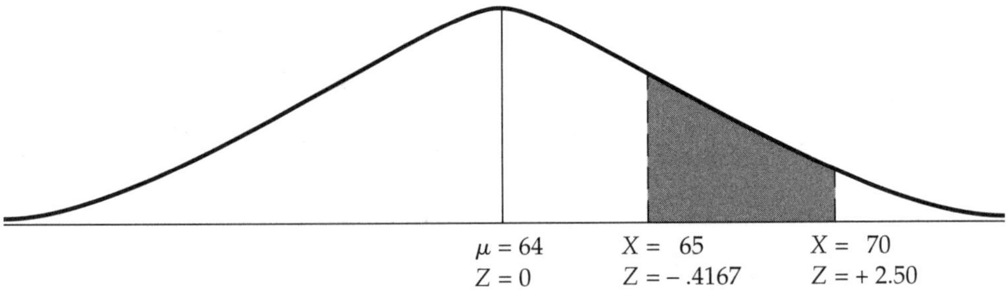

point to the left of the distribution is seen as 0%, whereas the furthest right point is seen as 100%.

To clarify the discussion thus far it is helpful to use an actual example. Let's say we want to know the height that 25% of all women are equal to or shorter than. Utilizing the above formula, we already know the mean is 64 inches and the standard deviation is 2.4 inches. As such, the only missing piece of information in this formula is the Z score. To make this determination, it is helpful to once again visualize the problem applied to the normal curve. This is Step 1. Since 25% is less than 50%, it is placed to the left of the mean and the corresponding area below it is shaded in (see Figure 6.9).

Next, we need to decide which of the three columns of probability scores we should use. Since we are concerned with the shaded region, which is "The Small Part" of the distribution, we look to column 4 and select the probability score that is closest to 25% (.25). The value of .25 falls between two values: .2514 and .2483. Of these two values the value that is closer to .25 is .2514. This is Step 2. The corresponding Z score for this value is $-.67$. This is Step 3. (Remember that all Z values to the left of the mean are negative. This is very important because failure to take a Z score's sign into account with this formula results in an incorrect answer.) The final step, 4, is to plug this Z score into the formula and do the appropriate calculations. These are summarized step-by-step below.

$$X = (-.67)\,2.4 + 64 = -1.608 + 64 = \boxed{62.392 \text{ inches}} = X$$

The final answer tells us that 25% of all women in the US are 62.392 inches or shorter. (Alternatively, we can also conclude that 75% of women in the US are 62.392 inches or taller.) Here is a summary of the four steps used to make this determination.

Step 1: Make a graphic representation with the information given and shade in the appropriate area under the normal distribution.

Step 2: Using this information, locate the percentage value in the appropriate column (always column 3 or 4) in Appendix 1.

FIGURE 6.9 25% Equal to or Shorter than

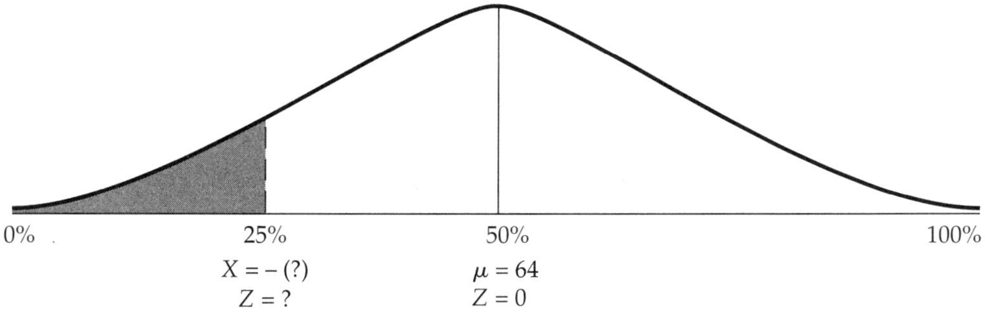

Step 3: Take the corresponding Z score from Step 2 and plug it into the formula with other given information.

Step 4: Calculate final score, which is your X of interest.

Here's another example. Let's say we now want to know what height 90% of women in the US are equal to or less than. Following the steps outlined above, we start by constructing a graphic representation of this information as it appears on the normal curve: Figure 6.10.

We then take this information and decide which column to use in Appendix 1. This is obviously column 3, "The Big Part" column. The percentage score closest to 90% (.90) is .8997. The corresponding Z score is +1.28 and, since 90% occurs on the right-hand side of the distribution, the corresponding Z score is positive. To calculate the final answer, we take this Z score and plug it into the formula. This is summarized below.

$$X = (1.28)\,2.4 + 64 = 3.072 + 64 = \boxed{67.072 \text{ inches}} = X$$

The final answer tells us that 90% of women in the US are 67.072 inches or shorter. That is, the vast majority of women in our society are five foot seven or shorter.

Figure 6.11 is an excellent way to visually summarize both of these operations: determining a probability value from an X value or determining the X value for a probability value. Until you become comfortable with making these determinations, we suggest that you refer to it.

PERCENTILES AND Z SCORES

As previously mentioned, we can also employ Z scores to calculate percentile rankings. Since a Z distribution by design ranks scores from lowest to highest, we can also use it to determine any observation's percentile ranking. And although the manner in which we calculate rankings is different than the previously discussed percentile-ranking formula, we still express percentile rankings

FIGURE 6.10 90% Equal to or Shorter than

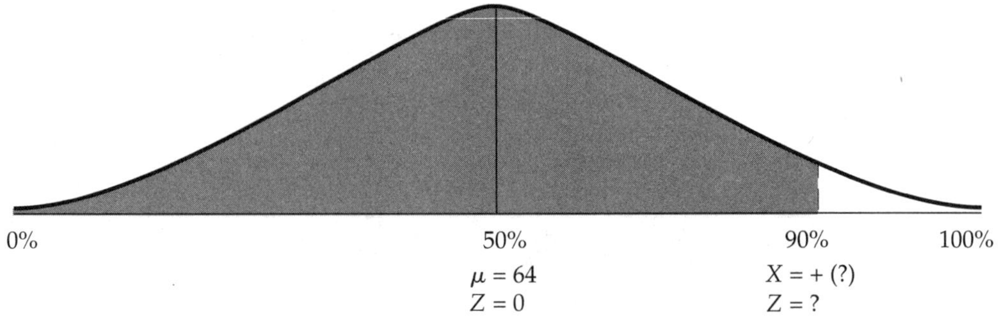

FIGURE 6.11 Summary of Steps Used to Determine Probability Values and Actual Scores Using a Normal Distribution

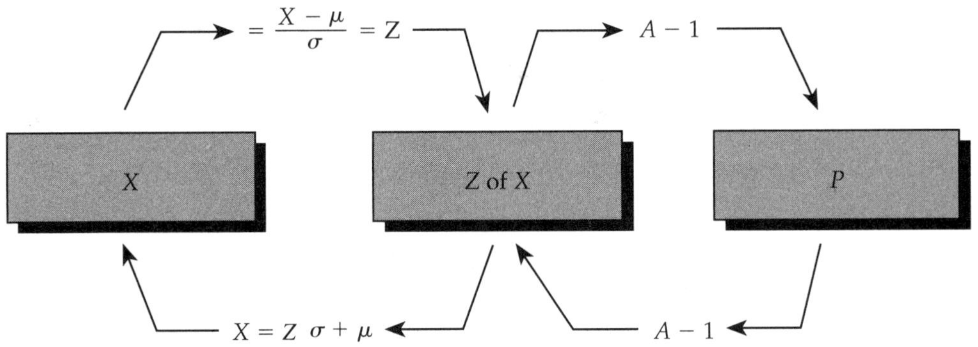

exactly as we did before (e.g., 22nd or 78th percentile ranking). Moreover, the Z distribution also enables us to determine any percentile ranking's actual score (X).

To do this we must view the Z distribution in the same manner as we did when calculating X's from percentages. That is, percentile rankings on a Z distribution are conceptualized in ascending order from lowest to highest (left to right) on the distribution. Thus, observations with the lowest percentile rankings (e.g., 2nd, 3rd, and 7th) are found in the far left tail region of the distribution; observations in the middle rankings (e.g., 47th, 50th, and 52nd) are found very near or on the mean of the distribution; and observations in the highest percentile rankings (e.g., 94th, 96th, and 99th) are found in the far right tail region of the distribution.

To clarify the discussion thus far and outline the steps involved in this application, let's go ahead and calculate some percentile rankings using Z scores. Instead of using heights, we have replicated the information on IQ scores below for the following example. (Note: Heights can just as easily be ranked.)

Average IQ Score for 4th-Grade Children: $\mu = 100$
Standard Deviation for 4th-Grade Children: $\sigma = 15$

To begin with, let's say we want to know the percentile ranking of a child that has an IQ of 80 (obviously a child with this low of an IQ has probably been over a cliff with Calvin one too many times). Like before, the first step is to calculate a Z score. This is done below.

$$\frac{80 - 100}{15} = \boxed{-1.333\bar{3}} = Z \text{ of } X \text{ or } Z \text{ Score for an IQ of } 80$$

The next step, exactly as before, is to make a graphic representation of this information applied to the normal distribution (see Figure 6.12). (Please note: Whenever we calculate a percentile ranking using Z scores we always shade in the area below the

FIGURE 6.12 Percentile Ranking for an IQ Score of 80

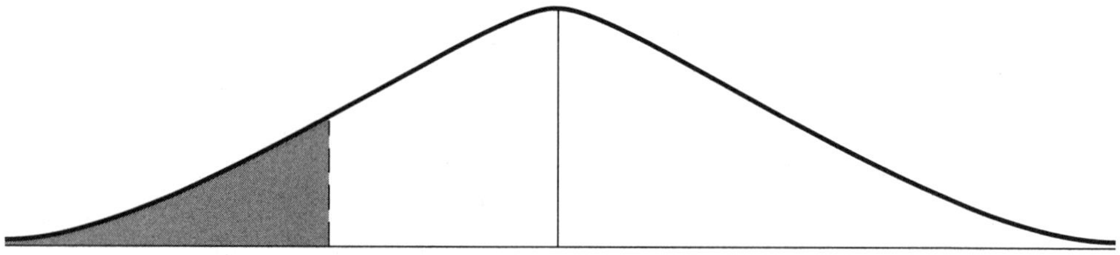

Z of X regardless of whether it is located above or below the mean. The reason for this, like the previously discussed formula used to determine percentile rankings, is that we are always concerned with the number of scores that fall *below* the X of interest relative to all the scores in the distribution.)

As before, the next step is to locate the calculated Z score (1.33) in Appendix 1. Having done this, we must next determine which column to utilize. At this point, there are three rules that should always be followed when determining percentile rankings: (1) Never use column 2, (2) always use column 3 when the Z score is positive (the X of interest is greater than the mean), and (3) always use column 4 when the Z score is negative (the X of interest is less than the mean). We never use column 2 because the area from mean to the Z score is not of interest when we calculate percentile rankings. Applying these rules to this example finds that we use column 4. This gives an answer of .0918. The final step is simply to move the decimal point over two places and to express this answer as a percentile ranking. Thus, a child with an IQ of 80 is found in the 9.18th percentile. The five steps just used in this application are summarized below.

Step 1: Calculate the Z score.

Step 2: Make a graphic representation of the Z of X in terms of where it is located on the normal curve and always shade in the area below—to the left—of the score.

Step 3: Locate your Z score in Appendix 1.

Step 4: Using the above information, determine the appropriate column (3 or 4) and record this answer.

Step 5: Take the answer from Step 4 and move the decimal point over two places. Expressed as a percentile ranking, this is your final answer.

Here's another example. Let's say we are now interested in the percentile ranking of a child with an IQ of 125. First we calculate a Z score (below).

$$\frac{125 - 100}{15} = \boxed{-1.666\overline{6}} = \text{Z of X or Z Score for an IQ of 125}$$

FIGURE 6.13 Percentile Ranking for an IQ Score of 125

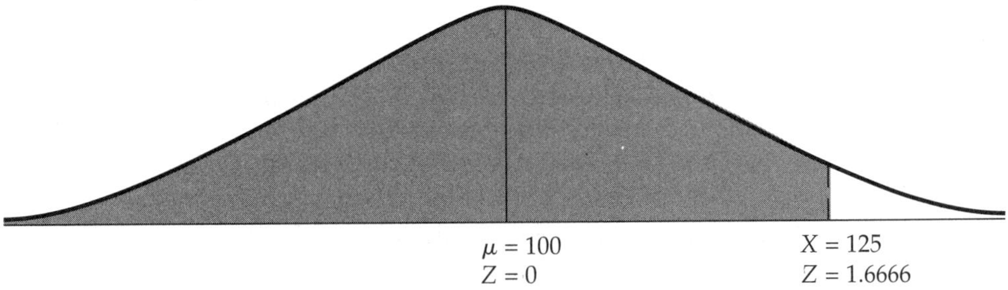

Following the steps, we next make a graphic representation of the normal distribution and shade in the area below the X of interest (see Figure 6.13).

Next, we locate the Z score (1.66) in the appendix, and find that the corresponding answer in column 3 is .9515. Finally, we move the decimal point over two places and our final answer is the 95.15th percentile. This tells us that a child with an IQ of 125 is ranked in the 95.15 percentile. In other words, this child is quite intelligent relative to other children since very few children have a higher IQ.

Of course, we could also take percentile rankings and convert them into actual scores. To accomplish this we follow the same steps presented in the section immediately preceding this one on converting probability scores into actual scores (X's). Since we have already done these operations, we will not redo them here.

Chapter Summary and Conclusions

This chapter investigated various ways of locating scores in a distribution relative to all the other scores in it. To this end, we first discussed percentile rankings which are calculated using actual raw scores that are ranked. Next, we described how Z scores applied to a normal distribution enable us to determine the percentage of scores that occur above, below, or between X's of interest. We also reversed the steps in this procedure and determined actual scores (X's) from percentages in the given distribution. Finally, we discussed how Z scores are used to determine percentile rankings of any observation of interest.

Key Terms To Remember

Percentile Rankings	"Z to the Mean" Column
Normal Distribution	"The Small Part" Column
Z Scores	"The Big Part" Column

PRACTICE EXERCISES

1. Let's go back to Cartoon 6.1 and our video game players. Instead of being interested in the video game scores of our 10-year olds, we are now concerned with their parents' yearly incomes. Each child's parents yearly income, expressed in thousands of dollars per year, is reported in Table 6.4. To simplify our presentation, we have continued to use the children's names to represent their parents' incomes.
   a. What is Nicole's percentile ranking?
   b. What is Harry's percentile ranking?
   c. What is Dan's percentile ranking?
   d. What is Helen's percentile ranking?
   e. What is Steve's percentile ranking?
2. Given the following information:
   IQ Scores: Mean (μ) = 100; Standard Deviation (σ) = 15
   What is the probability that the next person you will meet will have an IQ of
   a. 115 or greater?
   b. 100 or less:
   c. between 90 and 105?
   d. 130 or greater?
   e. 90 or less?
   f. between 110 and 120?
   (All of these questions can also be expressed in terms of what percentage of the population has an IQ of: _____.)
3. Using the above IQ information, what IQ score is
   a. 90% of the population equal to or less than?
   b. 35% of the population equal to or less than?
   c. 50% of the population equal to or less than?
   d. 15% of the population equal to or less than?
   e. 73% of the population equal to or less than?
4. Given the following information:
   Heights of Women in the US:
   Mean (μ) = 64; Standard Deviation (σ) = 2.4

TABLE 6.4 Parents' Yearly Incomes*

Incomes

Janice	30	Anna	42	Allan	22	Pina	60
Harry	33	Randi	69	Alex	34	Katie	66
John	36	Dan	53	Nicole	29	Billy	46
Jenny	37	Jeff	49	Roby	57	Steve	38
Jim	25	Chris	43	Helen	41	Bob	23

*Remember that your data set needs to be ranked-ordered.

What is the probability that the next woman you meet will be

a. 60 inches or shorter?
b. 70 inches or taller?
c. Between 61 and 58 inches?
d. 62 inches or taller?
e. 58 inches or shorter?
f. between 66 and 68 inches?

5. What height is it that
   a. 75% of all American women are equal to or shorter than?
   b. 86% of all American women are equal to or shorter than?
   c. 22% of all American women are equal to or shorter than?
   d. 37% of all American women are equal to or shorter than?
   e. 50% of all American women are equal to or shorter than?
6. Determine the percentile rankings of the following Z scores.
   a. –.75
   b. +1.5
   c. –1.5
   d. +3.0
7. Reusing the above information on women's heights, what is the percentile ranking of a woman who is
   a. 61 inches tall?
   b. 64 inches tall?
   c. 62.5 inches tall?
   d. 67 inches tall?
   e. 66.5 inches tall?
   f. 69 inches tall?

Chapter 7

Probability

Many students immediate reaction to this chapter is, "I hate probabilities," and thus, "This is going to be one of the most difficult chapters." We also shared this adverse attitude about probabilities when we were presented with this material in our first undergraduate statistics course. The following Bloom County cartoon perhaps best summarizes these attitudes. Assuming that you have read the chapters in the order of their presentation, perhaps some of you thought there wasn't a very good chance ("probability") that you would make it this far into a statistics text. Isn't it amazing what Opus and a good instructor can accomplish?!

Having both personally survived presentations on probability as students and observed our own students "survive" and master the information found in this chapter, we have discovered over the years that it is no more difficult than any of the other presentations found in this text. Moreover, we have learned of its importance in theoretically understanding Chapter 6 and the chapters that follow. Further, we have included only the material that we see as vital in understanding basic probability theory. On a more practical level, those of you who have been contributors to games of chance ("chance" here being synonymous with probability) in Las Vegas and other gambling venues may be interested in potential techniques to increase your winnings or, more realistically, to decrease your losses.

To a certain extent, all of us are probability theorists. Daily, we ask ourselves questions such as:

- What is the probability that I can run this red light and not get a ticket or have an accident?
- What is the probability that I can pass the next quiz if I don't study for it?
- What is the probability that I can miss my date with Lee tonight and she/he will still speak to me tomorrow?

In other words, all of us constantly wonder about the probability of certain events occurring in our lives.

CARTOON 7.1

On a more scientific level, our lives are even more directly influenced by probabilities. Weather forecasters tell us what type of weather we can expect tomorrow in terms of probabilities; a forecast for a 60% chance of rain tells us that we might want to take an umbrella with us on our travels. Actuaries, people who calculate insurance rates, use probabilities to dictate how much the monthly premiums are for our life, medical, and car insurance. The quality and safety of most consumer products are also determined using probabilities.

Furthermore, as we will discover in the following chapters, there are many statistical techniques that enable social scientists to calculate the probability that they have arrived at a correct conclusion in their research. Some of these statistical techniques are the same ones that enable pollsters to infer political attitudes of the entire United States population and to predict the winners of upcoming elections using just a sample. In sum, although we may not recognize it, in reality a great deal of the quality of our existence is determined by probabilities. As such, like statistics in general, there is utility in all of us understanding probabilities.

This chapter begins with a discussion about basic rules of calculating probabilities for independent events, for "and" statements, and for "or" statements. This is followed by a discussion of how permutations and combinations are used to determine the probability of certain outcomes. The chapter ends with a demonstration of how combinations can be used to construct a frequency distribution of probabilities on which the normal curve (discussed in Chapter 6) is theoretically based upon.

BASIC PROBABILITIES: INDEPENDENT EVENTS, "AND," AND "OR" STATEMENTS

When asked simple questions, like "What is the probability of getting heads when flipping a balanced coin?" or "What is the probability of a child's being born female?," we would guess that most of you intuitively know what both answers are: 50%. As an aside, we should note that blemished, unbalanced coins can result in the coin's consistently coming up heads or tails, while the probability of a woman's giving birth to a female child is actually slightly less than 50%. The following examples will assume that one is using a balanced coin and that there actually is a 50/50 chance of having a female or male child.

We would also guess that, mathematically speaking, many of you may not know how this outcome is determined. Quite simply, this is accomplished by dividing the number of *desired independent* outcomes, the numerator, by the number of *possible independent* outcomes. In both these examples, and the ones that follow, an outcome is considered independent if it can occur only one way out of two or more possible ways (e.g., heads or tails, male or female). As such, it can also be referred to as a discrete or mutually exclusive outcome (event).

In the case of flipping a balanced coin and getting a head, it is mathematically expressed as:

Desired outcome of a head ⟶ 1
Number of possible outcomes (heads/tails) ⟶ 2

$$\frac{1}{2} = \boxed{.50 \text{ or } 50\% \text{ chance}}$$

Alternatively, and using Cartoon 7.2 as a backdrop, one can similarly calculate the probability of a woman's having a female child (with Calvin as an example, there are obviously more than enough male children running around already).

Desired outcome of a female ⟶ 1
Number of possible outcomes (male or female) ⟶ 2

$$\frac{1}{2} = \boxed{.50 \text{ or } 50\% \text{ chance}}$$

CALVIN AND HOBBES By Bill Watterson CARTOON 7.2

Applying this same logic to other events that have independent outcomes, we can ask questions such as "What is the probability of getting a five on one roll of a balanced die?"; "What is the probability of getting the number 27 on one spin of a roulette wheel?"; or "What is the probability of getting a red number on one spin of a roulette wheel?" Taking each of these in turn, the die question is answered by simply deducing that we are concerned with one outcome of an event that can occur six possible ways—the six sides/numbers on a die. This is mathematically expressed as follows.

Desired outcome of a five ⟶ / Number of possible outcomes on a die ⟶ $\frac{1}{6} = \boxed{.166\bar{6} \text{ or } 16.66\% \text{ chance}}$

Thus we conclude that there is a 16.66% chance of us getting a five on one roll of a die. Moreover, we can also conclude that the probability of any one of the numbers (sides) occurring is also a 16.66% chance.

Turning to the roulette wheel question, we must first determine the number of possible outcomes that can occur on one spin. Hypothetically, let's say that the wheel under consideration has 100 numbers (1–100) on it of which 1/2 are black and 1/2 are red (50 black and 50 red). This information allows us to ascertain that in the first question we are concerned with one possible outcome, a 27, out of 100 possible outcomes. Or, as mathematically expressed below, there is a 1% chance of getting a 27 on one spin of a roulette wheel.

Desired outcome of a 27 ⟶ / Number of possible outcomes (numbers) on roulette wheel ⟶ $\frac{1}{100} = \boxed{.01 \text{ or } 1\% \text{ chance}}$

Turning to the second roulette question, the probability of getting a red number, the answer is determined by dividing the number of red spaces on the wheel, 50, by the overall number of spaces found on it, 100. Thus, as expressed below, there is a 50% chance of a red number's occurring with any one spin. In other words, the probability of getting a red number is much higher than getting any one number on a single spin of the roulette wheel.

Desired outcome of a red number	⟶	50		
Number of possible outcomes (numbers) on roulette wheel	⟶	100	=	.50 or 50% chance

Box 8

With events such as flips of a coin, the shake of a die, or the spinning of a roulette wheel, many people often falsely believe that the occurrence of one outcome diminishes the likelihood of its subsequent occurrence. For instance, if one flips a coin and desires heads, there is a 50% chance of this event's occurring. Assuming it does come up heads, using the above incorrect logic one might surmise that there is a decreased probability of its being heads on the next consecutive flip. This is simply not true. Since each time the coin is flipped it is still an independent event and not contingent upon the occurrence of events that have preceded or that will potentially follow, the probability of heads on the next flip is still 50%. The logic of inferring a decreased probability of occurrence in events such as a flip of a coin is known as *The Gambler's Fallacy*. As demonstrated shortly, however, we can calculate the probability of certain events, such as heads on a coin or a male child's being born, occurring in a certain order.

Another slightly more involved example that we can utilize is a deck of cards. For you non-card players, we will briefly review what possibilities—types of cards—are found in a deck of cards. To begin with, a deck is made up of 52 cards: the overall number of possible outcomes. Further, each deck has four suits—Hearts, Diamonds, Spades, and Clubs—of 13 cards each. Within each suit is an ace (4 aces in a deck); three face cards—a King, a Queen, and a Jack (12 face cards in a deck); and non-face cards that represent the values of two through ten (36 of these cards in a deck). (Please note: For simplicity we do not consider the two joker cards here or in subsequent examples.)

Knowing this, we can now ask simple questions such as "What is the probability (how much of a chance is there) of drawing an ace from a deck of shuffled cards?" (For this and all the following card examples to work, we have to assume theoretically that the deck of cards has been shuffled several times. This allows us to assume that the cards are randomly located throughout the deck, and, as such, that each card has an equal chance—possibility—of being selected.) Using the same logic used to mathematically determine the likelihood of flipping a head on a coin or giving birth to a female child, there are two pieces of information we must first determine: (1) the number of desired outcomes, to be divided by (2) the number of possible outcomes. The number of desired outcomes (an ace) is four, which is divided by the number of possible outcomes, in this case 52 (there are 52 cards in a deck that one could possibility select). All of this is mathematically presented below. From this answer we conclude that with one draw there is a 7.69% (almost an 8 in 100) chance of drawing an ace out of a deck of shuffled cards.

$$\frac{4}{52} = \frac{1}{13} = \boxed{.0769 \text{ or a } 7.69\% \text{ chance}}$$

Using this same logic, we can also ask "What is the probability of drawing a face card from a deck of shuffled cards?" As we have already noted, there are 12 face cards, the number of desired outcomes, which is divided by the number of all possible outcomes, 52. As completed below, we find that there is a slightly greater than 23% chance (nearly 1 out of 4) of drawing a face card from a deck of shuffled cards. Conversely, however, there is still a greater chance of not drawing a face card (76.92%).

$$\frac{12}{52} = \frac{3}{13} = \boxed{.2308 \text{ or a } 23.08\% \text{ chance}}$$

All of the calculations we have performed so far have been fairly simple and in all "probability" some of you may have already known how to do them. However, a slightly more difficult set of operations is required to answer questions such as "What is the probability of drawing a red ace *or* a black 5 from a deck of shuffled cards?" In a sense, we are now faced with two sets of desired outcomes connected with an "*or*" statement. As many of you may have guessed, an "or" statement, as found in the title of this section, involves a further mathematical operation. Specifically, when an "or" statement is encountered, we are instructed to add together different probability events.

To this end, we first calculate the probability of each separate outcome, and then simply add them together. In other words, first we calculate the probability of drawing a red ace out of a deck of shuffled cards (2/52) and the probability of drawing a black 5 out of a deck of shuffled cards (2/52). Then we simply add these two values together to get the final answer. One way we can add these values together is to use the separate probability values expressed as fractions. Since, however, the present example has a common denominator of 52, the numerator values are simply added together to get the final answer. (Note: Whenever fractions are added together, they must have a common denominator. If they do not, the final answer will be incorrect.) All of this is mathematically done below.

$$\frac{2}{52} + \frac{2}{52} = \frac{4}{52} = \frac{1}{13} = \boxed{.0769 \text{ or a } 7.69\% \text{ chance}}$$

Turning to another example, we now want to know "What is the probability of drawing the Queen of Hearts *or* a 4 from a deck of shuffled cards?" Once again, we independently determine the probability of each potential outcome and then simply add these values together to get the final answer. This is presented below.

$$\frac{1}{52} + \frac{4}{52} = \frac{5}{52} = \boxed{.0961 \text{ or a } 9.61\% \text{ chance}}$$

Considering yet another probability event, let's say that we are now interested in a die. Once again using an "or" statement, we want to know "What is the probability of rolling a 2 *or* a 5 on a balanced die?" As with previous examples, we first must determine the number of the desired outcomes, which is then divided by the overall number of possible outcomes. A die has six sides with six corresponding numbers; thus, there are six possible outcomes. Since two of the outcomes are desired, a 2 *or* a 5, we individually express them and then add these values together to get the final answer. Thus, in mathematical terms we have a 1 in 3 chance (33.33%) of getting a 2 or a 3 on one shake of a die.

$$\frac{1}{6} + \frac{1}{6} = \frac{2}{6} = \boxed{.333\bar{3} \text{ or a } 33.33\% \text{ chance}}$$

To most of you, what has been presented thus far probably still seems quite simple. However, what if the above question was changed ever so slightly and instead read "What is the probability of rolling a 2 *and* then a 5 on a balanced die?" In other words, not only is the die shaken two separate times but, instead of an "or" statement, we are now presented with an "and" statement. This, of course, means that a different mathematical operation is used. Specifically, where an "or " statement informs us to *add*, an "and" statement tells us to *multiply*.

With this restated question, we are now required to multiply two separate fractions together to obtain the final answer. Since the probability of independently rolling a 2 on a balanced die is 1 out of 6, as is the probability of independently rolling a 5 (out of 6), we multiply these two values together to get the final answer. Thus, the probability of rolling a 2 *and* then a 5 on a balanced die in this exact order is a rather unlikely outcome with a 2.77% chance of occurrence. This is mathematically demonstrated below.

$$\frac{1}{6} \times \frac{1}{6} = \frac{1}{36} = \boxed{.027\bar{7} \text{ or a } 2.77\% \text{ chance}}$$

To clarify the above calculations, when fraction values are multiplied we first multiply one numerator by the other numerator and, correspondingly, one denominator by the other denominator. To obtain the final answer we simply divide the resultant fraction's numerator by its denominator. This is how the above was completed. (As an aside, an alternative solution to the above problem is to separately divide each fraction and then multiply these two values together to obtain the final answer. For the above, this is $.1666 \times .1666 = .0277$.)

We now return to our deck of cards, with its much larger number of possible outcomes, to create another example of an "and" statement. Let's say we have two decks of shuffled cards from which we want to know the probability of drawing a King from the first deck *and* then a Spade from the second deck. Using the proper procedural steps, we first determine the probability of each separate event *and* then multiply these two values together to obtain the final answer. That is, there are four Kings and 13 Spades, and thus, as expressed below, there is a 1.92% chance of drawing a King *and* then a Spade from two separate decks of cards.

$$\frac{4}{52} \times \frac{13}{52} = \frac{52}{2704} = \boxed{.0192 \text{ or a } 1.92\% \text{ chance}}$$

Using our cards for one last example of this type of operation, let's say we are interested in the probability of drawing a one-eyed Jack *and* then an ace. Since there are two one-eyed Jacks and four aces, we simply multiply these two values together and then divide the resultant numerator value by its corresponding denominator value. These steps and the final answer are presented below. In sum, there is less than a 1% chance (.29%) of drawing a one-eyed Jack *and* then an ace. Some of you might keep this in mind the next time you are playing Black Jack, a game of chance where an ace and a Jack are highly sought-after cards.

$$\frac{2}{52} \times \frac{4}{52} = \frac{8}{2704} = \boxed{.0029 \text{ or a } .29\% \text{ chance}}$$

Finally, let's change our example back to coins and ask the following question: "What is the probability of getting three heads in a row out of three consecutive flips of a coin?" To make this question more applicable in terms of operations we are presently reviewing, it can be restated more clearly to read "What is the probability of flipping a head *and* then a head *and* then a head?" The two "and" statements tell us to multiply the three separate probability values together. That is, since each of these outcomes has an independent 50% chance of occurring, we simply multiply 1/2 (or 0.50) together three times. Both of these identical operations are done below. Thus, there is a 12.5% chance of getting three heads in a row.

$$\frac{1}{2} \times \frac{1}{2} \times \frac{1}{2} = \frac{1}{8} = \boxed{.125 \text{ or a } 12.5\% \text{ chance}}$$

$$\text{Or} \quad .5 \times .5 \times .5 = \boxed{.125 \text{ or a } 12.5\% \text{ chance}}$$

As a noteworthy aside, this question reflects our previous discussion on the Gambler's Fallacy. That is, although there is only a 12.5% chance of getting three heads in a row, the probability of each independent event's (a head) subsequent occurrence is still 50%.

Factorials, Permutations, and Combinations

While "and" and "or" statements are important operations in determining probability events, their application is rather limited to very simple examples. To consider more complex problems we use what are called permutations and combinations. Before we explain what permutations and combinations are and how they are calculated, a brief discussion of factorials is warranted (this may be a review for some of you). The basic symbol that represents a factorial is an exclamation mark (!). Mathematically, this tells us to take whatever number is given and simply multiply it by every descending number found below it in the order of their occurrence to the value of 1. Applying this to an example, let's say we are asked to calculate 4! (stated four factorial). The final answer, 24, and the steps required to derive this answer are presented below.

$$4! = 4 \times 3 \times 2 \times 1 = \boxed{24}$$

or

$$(4 \times 3 = 12) \times 2 = 24 \times 1 = \boxed{24}$$

Or, alternatively, the required calculations and answers for 10!, 7!, 5!, and 3! are all found below.

$$10! = 10 \times 9 \times 8 \times 7 \times 6 \times 5 \times 4 \times 3 \times 2 \times 1 = \boxed{3{,}628{,}800}$$

$$7! = 7 \times 6 \times 5 \times 4 \times 3 \times 2 \times 1 = \boxed{5{,}040}$$

$$5! = 5 \times 4 \times 3 \times 2 \times 1 = \boxed{120}$$

$$3! = 3 \times 2 \times 1 = \boxed{6}$$

[Some of you may have a calculator with a factorial (!) button on it. To utilize this button (on most calculators) you simply press the desired number, press enter, and then press the factorial button.]

One quickly notices that factorials for even fairly small numbers (e.g., 10) result in rather large answers. Since many calculators, such as the one you may own, only go to 8 or 10 decimal points, this presents problems when we are calculating larger factorials. Since we are required to divide factorial values using both permutations and combinations, a simple way around the potential problem of large factorials is to write out each factorial. Then we simply cancel out all of the common numerator and denominator values. For instance, let's say we are asked to divide 12! by 10!. As the above suggests, we can do the calculations for this question two different ways and still get the same answer. Each of these ways is separately completed below. Since our students have overwhelmingly preferred the second approach of canceling out common values, we use it in all of the proceeding examples.

$$\frac{12!}{10!} = \frac{12 \times 11 \times 10 \times 9 \times 8 \times 7 \times 6 \times 5 \times 4 \times 3 \times 2 \times 1}{10 \times 9 \times 8 \times 7 \times 6 \times 5 \times 4 \times 3 \times 2 \times 1} = \frac{479{,}001{,}600}{3{,}628{,}800}$$

$$= \boxed{132}$$

versus

$$\frac{12!}{10!} = \frac{12 \times 11 \times \cancel{10 \times 9 \times 8 \times 7 \times 6 \times 5 \times 4 \times 3 \times 2 \times 1}}{\cancel{10 \times 9 \times 8 \times 7 \times 6 \times 5 \times 4 \times 3 \times 2 \times 1}} = \frac{12 \times 11}{1}$$

$$= \frac{132}{1} = \boxed{132}$$

We need to make note that when factorial values are canceled out, since we cannot have a zero (0) factorial, its value always becomes one (1). Applying this rule to the above example, as demonstrated, when all the numbers in the denominator are canceled out, we are still left with the value of one (1). Moreover, even in instances where an actual zero factorial (0!) is presented, since you cannot have a zero factorial, the value of one is still used. The reason for this rule becomes more apparent shortly.

Beginning with factorials applied to permutations, this type of operation, similar to an "and" statement, is always concerned with an outcome's occurring in a specific order for a given event. Permutations are also calculated using the assumption that once an element has occurred, it is no longer considered in future calculations; this is referred to as *without replacement.*

To simply demonstrate what "without replacement" means, we first refer back to an "and" statement. Let's say we have four cards, with each card representing one of the four suits. Using this information, we want to know "What is the probability of drawing a Heart *and* then a Spade?" On the first draw there is a 1 out of 4 chance of getting a Heart. Then, for the second draw, there is a 1 out of 3 chance of getting a Spade (note that we have *not replaced* the first card back into the deck). Finally, the two resultant fractions are simply multiplied together. This is presented mathematically below. Thus, we can conclude that there is a 1/12 or 8.33% chance of drawing a heart *and* then a spade.

$$\frac{1}{4} \times \frac{1}{3} = \frac{1}{12} = \boxed{.083\bar{3} \text{ or a } 8.33\% \text{ chance}}$$

An alternative, perhaps easier, way we can obtain this same answer is to use the following permutation formula. Four different symbols are found in this formula: (1) *nPm* simply represents the formula for permutations, (2) *N* is the total number of elements under consideration, (3) *M* is the number of elements considered in a specific order, and (4) "!" simply means that the given numerical value is treated as a factorial.

$$nPm = \frac{N!}{(N - M)!}$$

N = total number of elements considered
M = number of elements considered in a specific order

To apply our card example to this formula, we first determine that the total number of elements (cards) is 4; thus, $N = 4$. Then we determine that the number of elements considered in a specific order is 2, a Heart *and* then a Spade; thus, $M = 2$. All of this is summarized below, with the resulting answer of 12. [Note: Since the $(N-M)$ is in parentheses, using the rules of algebraic order we must first do the operation within it $(4 - 2)$ and then the factorial.]

$$nPm = \frac{4!}{(4-2)!} = \frac{4 \times 3 \times 2 \times 1}{2 \times 1} = \frac{4 \times 3}{1} = \boxed{12}$$

Twelve is obviously not the final answer. The answer of 12 tell us that there are 12 different ways that two cards out of four can occur in a specific order. Since we are only interested with one of these outcomes, a Heart *and* then a Spade, we simply divide 1 by 12 (1/12), which once again gives us the final answer of $.083\bar{3}$ or a 8.33% chance. Once again, this is mathematically presented below.

$$\text{Probability} = \frac{1}{P^*} = \frac{1}{12} = \boxed{.083\bar{3}} = \text{Or there is a 8.33\% chance of this outcome in this specific order}$$

* *P* represents the answer obtained from the permutation formula.

Let's change our example and now consider a 100-meter track race. We want to know "What is the probability of picking the top three finishers, in the correct order of their completing the race, out of six runners competing in the race?" We are assuming that the racers are nearly identical in their individual capabilities. In reality, when races are run, whether by people or horses (as in the next example), seldom if ever are the participants equally matched. This notwithstanding, we use these as examples of using permutations since most of you are familiar with racing events. Referring back to the permutation formula, we discern that $N = 6$ (there are six runners in this race) and that $M = 3$ (the top three finishers of the race in the exact order of finishing—1st, 2nd, and 3rd).

Below, we have plugged these values into the permutation formula and obtained the value of 120. Once again, the obtained value of 120 tells us that there are 120 different ways that the top three finishers of a race comprised of six runners can occur. Since we are only interested in one outcome—one race—the value of one (1) is divided by 120 (1/120); the resultant answer tells us that we have a .83% chance of picking the top three of six runners of a race.

$$nPm = \frac{6!}{(6-3)!} = \frac{6 \times 5 \times 4 \times 3 \times 2 \times 1}{3 \times 2 \times 1} = \frac{6 \times 5 \times 4}{1} = \boxed{120}$$

$$\text{Probability} = \frac{1}{P} = \frac{1}{120} = \boxed{.008\bar{3}}$$ = Or there is a .83% chance of this outcome occurring in this specific order

For the next example, let's shift our interests from track races to horse races. Let's say that we are interested in the probability of picking the top two finishers of a race with 10 horses in it. Referring back to the permutation formula, we first determine that $N = 10$ (total number of horses in the race) and $M = 2$ (the number of horses considered in the exact order of their finishing 1st and 2nd). These values are plugged into the formula to give us an answer of 90—there are 90 different ways that the top two finishers of a race with 10 horses in it can occur. Finally, since we are interested in one outcome—one race—we divide 1 by 90 (1/90), which gives us the final answer of $.011\bar{1}$. In other words, we have a 1.11% chance of picking the top two finishers of a race comprised of 10 horses.

$$nPm = \frac{10!}{(10-2)!} = \frac{10 \times 9 \times 8 \times 7 \times 6 \times 5 \times 4 \times 3 \times 2 \times 1}{8 \times 7 \times 6 \times 5 \times 4 \times 3 \times 2 \times 1} = \frac{10 \times 9}{1}$$

$$= \boxed{90}$$

$$\text{Probability} = \frac{1}{P} = \frac{1}{90} = \boxed{.011\bar{1}}$$ = Or there is a 1.11% chance of this outcome's occurring in this specific order

The operations required to calculate permutations are fairly straightforward. And though the answers that permutations give us are interesting, a related probability technique (called combinations) is far more valuable in understanding some of the underlying assumptions of statistics. Combinations allow us to answer questions such as "What is the probability of having two male children—hopefully not two children like Calvin—out of five consecutive births?" This is a far different question than the ones asked using permutations.

That is, combinations are not concerned with a specific order of an outcome, but rather the probability of an outcome without concern for order. Thus, in the above question we are not concerned with the order in which two male children may occur, but rather the probability that two male children out of five consecutive births will occur. Combining questions such as this with all other possible outcomes of the sex of a child (in the present case, out of five consecutive births) allows us to construct what is called a probability distribution (this is demonstrated shortly). This is where the real value of using combinations is found in understanding statistics.

Before actually turning to the combination formula and its use, we will first determine the overall number of possibilities associated with five consecutive births. This is accomplished by using a simple formula that considers the number of possibilities for each event and the overall number of events. Specifically, we use the numerical value of possible outcomes for each independent event to the power of the overall number of consecutive events under consideration. In the present example, this is 2 (there are two possible outcomes for each event: a male or female child) to the fifth power (five consecutive births). This is done below. The resultant answer tells us that there are 32 different ways—combinations—that male and/or female babies can occur out of five consecutive births. (Two to the fifth power simply means that we take $2 \times 2 = 4; 4 \times 2 = 8; 8 \times 2 = 16; 16 \times 2 = 32$).

$$2^5 = \boxed{32} \quad \text{(2 possible outcomes for 5 independent events—births)}$$

Next we must determine the number of desired outcomes: two male children out of five consecutive births. To demonstrate the utility of the combination formula, however, we first take a more laborious approach to answer this question. This is accomplished deductively by writing out each possible way two male births out of five consecutive births can occur (see Table 7.1). While we discover that there are 10 ways that two male children can occur out of five consecutive births, one should also note that writing out each possibility is not only time consuming but the distinct possibility for error exists. With a more complex question, such as the probability of five male children out of 10 consecutive births, these conclusions are even more relevant.

Nevertheless, the final answer for our initial question is arrived at by dividing the value for the number of ways two male children can occur (10) by the total number of possible ways five consecutive births can occur (32); this equals .3125. Therefore, there is a 31.25% chances that a woman will have two male children out of five consecutive births.

Desired outcome of two male children ⟶ 10
Number of possible outcomes for five consecutive births ⟶ 32

$$\frac{10}{32} = \boxed{.3125 \text{ or a } 31.25\% \text{ chance}}$$

TABLE 7.1 Possible Ways for Two Male Children Out of Five Consecutive Births

	Birth 1	*Birth 2*	*Birth 3*	*Birth 4*	*Birth 5*
1	M	M	F	F	F
2	M	F	M	F	F
3	M	F	F	M	F
4	M	F	F	F	M
5	F	M	M	F	F
6	F	M	F	M	F
7	F	M	F	F	M
8	F	F	M	M	F
9	F	F	M	F	M
10	F	F	F	M	M

Using the combination formula below, we can answer this same question in a far more expedient manner. Once again, the *nCm* symbolizes the formula for combinations; however, although the symbols of *N* and *M* are still in the formula, their meaning and application are slightly different—we now also consider an independent *M*!. As noted, *N* represents the number of consecutive events considered while *M* is the number of elements of interest considered.

$$nCm = \frac{N!}{M!\,(N - M)!}$$

N = number of consecutive events considered
M = number of elements of interest considered

Thus, in our present example, $N = 5$ (five consecutive births) while $M = 2$ (two male births). These values are placed into their appropriate places in the formula below to determine again that there are 10 possible ways that two male children out of five consecutive births can occur. Once again, we divide the value of 10 by the value of 32—the number of possible ways 5 consecutive births can occur—to give us the final answer of a 31.25% chance.

$$nCm = \frac{5!}{2!\,(5 - 2)!} = \frac{5 \times 4 \times 3 \times 2 \times 1}{2 \times 1\,(3 \times 2 \times 1)} = \frac{5 \times 4}{2 \times 1} = \frac{20}{2} = \boxed{10}$$

Overall number of possible combinations is $2^5 = \boxed{32}$

$$\frac{\text{Desired outcome of two male children}}{\text{Number of possible outcomes for five consecutive births}} \longrightarrow \frac{10}{32} = \boxed{.3125 \text{ or a } 31.25\% \text{ chance}}$$

For the next example, let's say we want to know the probability (the chance) of a mother having five male children out of seven consecutive births. Following the order of the steps used in the previous example, we first calculate the overall number of possible combinations (outcomes) associated with seven consecutive births. This is derived, once again, by simply taking the number of possible independent outcomes for each event to the power of the overall number of events under consideration. For the present example, as completed below, we take two to the seventh power to get the obtained value of 128. In other words, there are 128 different ways (combinations) that seven consecutive births can occur.

Overall number of possible combinations is $2^7 = \boxed{128}$

Next we must determine the number of possible ways that five male children out of seven consecutive births can occur. Below we have entered the appropriate values into the combination formula to learn that there are 21 different combinations (ways) in which five male children out seven consecutive births can occur. We divide the value of 21 (the number of desired outcomes) by the overall number of possible outcomes (128) and obtain the final answer of .1640. Thus, there is a 16.40% chance of five male children being born out of seven consecutive births.

$$nCm = \frac{7!}{5!\,(7-5)!} = \frac{7 \times 6 \times 5 \times 4 \times 3 \times 2 \times 1}{5 \times 4 \times 3 \times 2 \times 1\,(2 \times 1)} = \frac{7 \times 6}{2 \times 1} = \frac{42}{2} = \boxed{21}$$

Desired outcome of five male children ⟶
Number of possible outcomes for seven consecutive births ⟶ $\frac{21}{128} = \boxed{.1640 \text{ or a } 16.40\% \text{ chance}}$

For the next example, let's not only use a different event (flips of a coin), but also state the question in a different format: "What is the probability of flipping *three or fewer heads* out of eight consecutive flips of a coin?" This question is phrased in a manner nearly identical to that used for calculating probabilities using a Z-score distribution in Chapter 6, i.e., an IQ of 80 or less. However, since flips of a coin are discrete outcomes, heads or tails versus an endless array possible values associated with Z distributions (this distinction will be discussed in greater detail shortly), we determine the probability of this event using the combination formula. Further, as some of you may have already guessed, we have to use this formula more than once to answer the entire question.

Mathematically, the question is actually fourfold; that is, we are asking the probability of flipping three, two, one, and zero heads out of eight consecutive flips of a coin. Thus, we must make four separate combination calculations: (1) for the probability of getting three heads, (2) for the probability of getting two heads, (3) for the probability of getting one head, and (4) for the probability of getting zero heads (students often forget to calculate this last probability of zero heads). Then, since we are interested in the possibility of all of these outcomes happening, designated by the *or* statements, we add up each probability to get our final answer.

Before the formula calculations are undertaken, once again we first determine the overall number of ways that eight consecutive flips of a coin can occur. Since there are two possible outcomes for each independent event (a flip of a coin—heads or tails) and we are interested in eight consecutive events (flips), we simply take two to the eighth power. The resultant answer tells us there are 256 different ways (combinations) for eight consecutive flips to occur.

Overall number of possible combinations is $2^8 = \boxed{256}$

Next we need to independently calculate the combination values, each of which is then divided by the above value of 256. All of these calculations are completed as follows.

$$nCm = \frac{8!}{3!\,(8-3)!} = \frac{8 \times 7 \times 6 \times 5 \times 4 \times 3 \times 2 \times 1}{3 \times 2 \times 1\,(5 \times 4 \times 3 \times 2 \times 1)} = \frac{8 \times 7 \times 6}{3 \times 2 \times 1}$$

$$= \frac{336}{6} = \boxed{56} \text{ and thus } \frac{56}{256} = \boxed{.2188 \text{ or a } 21.88\% \text{ chance}}$$

and

$$nCm = \frac{8!}{2!\,(8-2)!} = \frac{8 \times 7 \times 6 \times 5 \times 4 \times 3 \times 2 \times 1}{2 \times 1\,(6 \times 5 \times 4 \times 3 \times 2 \times 1)} = \frac{8 \times 7}{2 \times 1}$$

$$= \frac{56}{2} = \boxed{28} \text{ and thus } \frac{28}{256} = \boxed{.1094 \text{ or a } 10.94\% \text{ chance}}$$

and

$$nCm = \frac{8!}{1!\,(8-1)!} = \frac{8 \times 7 \times 6 \times 5 \times 4 \times 3 \times 2 \times 1}{1 \times (7 \times 6 \times 5 \times 4 \times 3 \times 2 \times 1)} = \frac{8}{1} = \boxed{8}$$

$$\text{and thus } \frac{8}{256} = \boxed{.0312 \text{ or a } 3.12\% \text{ chance}^*}$$

and

$$nCm = \frac{8!}{^{**}\,0!\,(8-0)!} = \frac{8 \times 7 \times 6 \times 5 \times 4 \times 3 \times 2 \times 1}{1 \times (8 \times 7 \times 6 \times 5 \times 4 \times 3 \times 2 \times 1)} = \frac{1}{1} = \boxed{1}$$

$$\text{and thus } \frac{1^{***}}{256} = \boxed{.0039 \text{ or a } .39\% \text{ chance}}$$

* If one thinks about, it makes sense that there are eight ways in which a head can occur: one head for each of the eight consecutive flips of the coin with the other seven flips all being tails.

** As noted before, a 0 (zero) factorial is always equal to 1 (one).

*** This also makes logical sense. With eight flips of a coin, there is literally only one way that zero heads can occur: all tails.

Since there is an *or* statement in the original question, the final operation requires us to add together each of the probabilities to obtain an overall probability, the final answer.

$$.2188 + .1094 + .0312 + .0039 = \boxed{.3633 \text{ or a } 36.33\% \text{ chance}}$$

In other words, there is a 36.34% chance of flipping three heads or fewer out of eight consecutive flips of a coin.

Obviously, by using combinations in this manner we could answer an endless array of questions with reference to this or any other set of similar events. Let's go ahead and propose another question using heads and flips of a coin: "What is the probability of flipping four to six heads out of eight consecutive flips of a coin?" In other words, we are asking what the probability is of getting four *or* five *or* six heads out of eight consecutive flips of a coin.

Turning to the actual calculations, in the preceding example we determined that the overall number of possible outcomes for eight consecutive flips is 256. The same is true for this example; there are 256 possible outcomes. Next, we calculate separately each independent event of interest; thus, we calculate the probability of getting four heads, five heads, and six (four to six heads) out of eight consecutive flips. All of these calculations are done below.

$$nCm = \frac{8!}{4!\,(8-4)!} = \frac{8 \times 7 \times 6 \times 5 \times 4 \times 3 \times 2 \times 1}{4 \times 3 \times 2 \times 1\,(4 \times 3 \times 2 \times 1)} = \frac{8 \times 7 \times 6 \times 5}{4 \times 3 \times 2 \times 1}$$

$$= \frac{1680}{24} = \boxed{70} \text{ and thus } \frac{70}{256} = \boxed{.2734 \text{ or a } 27.34\% \text{ chance}}$$

and

$$nCm = \frac{8!}{5!\,(8-5)!} = \frac{8 \times 7 \times 6 \times 5 \times 4 \times 3 \times 2 \times 1}{5 \times 4 \times 3 \times 2 \times 1\,(3 \times 2 \times 1)} = \frac{8 \times 7 \times 6}{3 \times 2 \times 1}$$

$$= \frac{336}{6} = \boxed{56} \text{ and thus } \frac{56}{256} = \boxed{.2188 \text{ or a } 21.88\% \text{ chance}}$$

and

$$nCm = \frac{8!}{6!\,(8-6)!} = \frac{8 \times 7 \times 6 \times 5 \times 4 \times 3 \times 2 \times 1}{6 \times 5 \times 4 \times 3 \times 2 \times 1\ (2 \times 1)} = \frac{8 \times 7}{2 \times 1}$$

$$= \frac{56}{2} = \boxed{28} \text{ and thus } \frac{28}{256} = \boxed{.1094 \text{ or a } 10.94\% \text{ chance}}$$

To obtain the final answer, once again we sum the three separate answers together to inform us that there is a 60.16% chance of flipping four to six heads out of eight consecutive flips of a coin.

$$.2734 + .2188 + .1094 = \boxed{.6016 \text{ or a } 60.16\% \text{ chance}}$$

Probability Distributions for Combinations

Some of you may have noted a pattern emerging in the last two problems. That is, the calculation for five and six heads resulted in the same answers obtained for three and two heads. The reason that these answers are identical is that we are merely observing part of a probability distribution for a set of discrete events.

Flips of a coin or sex of a child are considered discrete events since they can only occur in fixed categories: heads or tails, male or female. Discrete events such as this are also reflective of a nominal level of measurement; per se, no numerical value can be directly applied to them. *Z* distributions, on the other hand, reflect continuous variables: weights, heights, income levels, etc. These types of variables represent a literally infinite number of possible of outcomes and, as such, they are considered an interval/ratio level of measurement. Nevertheless, we can apply the notion of a probability distribution to both types of variables. Since *Z*-score distributions and continuous variables have already been discussed, we now turn our attention to discrete probability distributions.

Reusing the example concerning eight flips of a coin to construct a probability distribution, most of the information needed has already been completed from previous examples. A probability distribution for eight flips of a coin must consider all possible outcomes, that is, for this example, all 256 possible combinations. One easy way that we can accomplish this task is to calculate the probability for each outcome (zero to eight heads), which can alternatively be viewed as zero to eight tails. Since we have already calculated the probabilities for zero to six heads, we can simply relist these values. To construct a complete distribution of combinations, however, we must also calculate the probabilities associated with seven and eight heads (as you will see, in a sense we have already done this by calculating the probabilities associated with one head and zero heads). Both of these calculations are completed below.

$$nCm = \frac{8!}{7!\,(8-7)!} = \frac{8 \times 7 \times 6 \times 5 \times 4 \times 3 \times 2 \times 1}{(7 \times 6 \times 5 \times 4 \times 3 \times 2 \times 1)\,1} = \frac{8}{1} = 8$$

$$\text{and thus } \frac{8}{256} - \boxed{.0312 \text{ or a } 3.12\% \text{ chance}}$$

and

$$nCm = \frac{8!}{8!\,(8-8)!} = \frac{8 \times 7 \times 6 \times 5 \times 4 \times 3 \times 2 \times 1}{(8 \times 7 \times 6 \times 5 \times 4 \times 3 \times 2 \times 1)\,1} = \frac{1}{1} = 1$$

$$\text{and thus } \frac{1}{256} = \boxed{.0039 \text{ or a } .39\% \text{ chance}}$$

We can now combine all eight of the calculations concerning eight consecutive flips (six from previous examples and two from the above example) to construct the probability distribution found in Table 7.2.

Several noteworthy observations can be made about this distribution. First, when all the combinations are added together they do in fact equal the number of all possible combinations: 256. Second, when all the probabilities of occurrence are added together they equal 1.00 or 100%. This should not be surprising since they do in fact represent *all* (100%) of the ways that eight flips of a coin can occur. Third, exactly like a *Z*-score distribution, the above distribution is symmetrical in shape. That is, other than its midpoint of four heads (like the mean in a *Z* distribution), all other values have an exact

TABLE 7.2 Discrete Probability Distribution For Eight Flips of a Coin

Number of Heads Out of Eight Flips	*Number of Combinations Out of Total Possible*	*Probability of Occurrence*
0	1/256	.0039
1	8/256	.0312
2	28/256	.1094
3	56/256	.2188
4	70/256	.2734
5	56/256	.2188
6	28/256	.1094
7	8/256	.0312
8	1/256	.0039
	Sum of combinations = 256	Probabilities = 1.00

corresponding value on the other side of the distribution. Moreover, just like the tail regions of the of Z-score distribution, the more one gets away from the midpoint value of four heads the more the probability of the given occurrence decreases. Finally, by graphing this information we can create an illustration (a histogram) that clearly represents the shape of a Z distribution: a bell-shaped curve. This is done in Figure 7.1, with the bell shape of a Z distribution superimposed on it.

Probability distributions, related to this one, are what make estimates of population parameters and hypothesis testing possible. The remainder of the text explores these important statistical concepts.

FIGURE 7.1 Discrete Probability Distribution

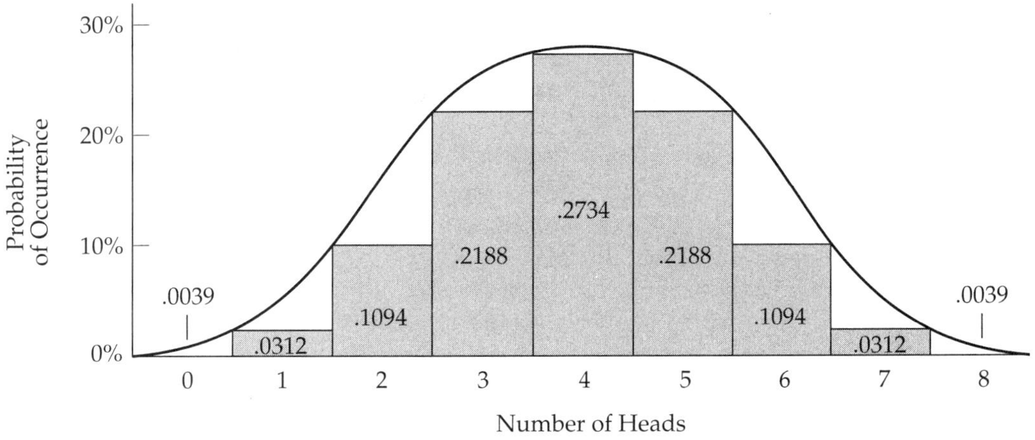

CHAPTER SUMMARY AND CONCLUSIONS

This chapter has explored several ways to determine the probability of certain events. Basic probabilities were discussed in terms of how they can solve problems involving independent events, "or" statements, and "and" statements. Examples used in these problems involved several different types of events from which probabilities can be determined (e.g., drawing a card, sex of a child at birth, flipping a coin). Next an overview of factorials was given and then applied to permutations and combinations. The latter, combinations, were demonstrated to have additional statistical meaning in that a probability distribution was constructed using them. Probabilities, although applied in a slightly different manner than in this chapter, will have increasing importance throughout the remainder of this text.

KEY TERMS TO REMEMBER

Basic Probabilities	Gambler's Fallacy
Independent Events	Factorials
"Or" Statements	Permutations
"And" Statements	Combinations

PRACTICE EXERCISES

1. a. Referring back to Cartoon 7.1, all things considered equal, "What was the probability of Opus's mother having a girl instead of Opus?"
   b. If Opus's mother wanted three (3) more children, "What is the probability of her having a girl, and then a boy, and then another girl?"
   c. If Opus was playing cards, "What is the probability of his randomly drawing (in a single draw) an ace or the Queen of Hearts out of a full deck of shuffled cards?"
   d. How about a black 4 or any face card (a King, a Queen, or a Jack)?
   e. If Opus was flipping a balanced coin, "What is the probability of his getting five heads in a row?"
   f. If he was shaking a balanced die and flipping a balanced coin, "What is the probability of his getting a 6 on the die and then a head on the coin?"
2. Using factorials with permutations:
   a. What is the probability of selecting a Club and then a Heart out of four (4) cards representing each suit (Clubs, Hearts, Spades, and Diamonds)?
   b. What is the probability of (randomly) selecting the top three (3) finishers of a bike race comprised of 7 bikers in the exact order of their finishing?
   c. What is the probability of (randomly) selecting the top two (2) finishers of a horse race comprised of 11 horses in the exact order of their finishing?

3. Using factorials with combinations:
   a. What is the probability of getting two heads out of seven flips of a balanced coin? (Obviously, order is no longer relevant.)
   b. What is the probability of having one female penguin out of 9 consecutive penguin births?
   c. What is the probability of having three or fewer male penguins out of 12 consecutive births?
   d. What is the probability of having five to eight female penguins out of 12 consecutive births?
   e. What is the probability of Opus's getting a date with one of these flightless female fowl? Obviously, given Opus's dating history and his rather sexist attitudes, the probability of his even making a successful first impression, let alone getting a date, is rather small. Nevertheless, per se, there is no definitive way to answer this question.

Chapter 8

Constructing Confidence Intervals

The last two chapters demonstrated techniques that determined the probabilities for continuous and discrete outcomes; e.g., there is a 50% chance the next person you will meet has an IQ of 100 or less or there is a 12.5% chance of flipping three consecutive heads. In this chapter, we take some of these basic ideas about probability theory and use them to construct what are called *confidence intervals.* Moreover, whereas the previous two chapters explored descriptive mathematical techniques, this chapter introduces the notion of *inferential statistics.* As such, this is the first chapter to delve exclusively into *sample statistics.*

During political campaigns the news media often make statements such as "If the election were held today, candidate A would have 52% of the votes while candidate B would have 47% of the votes." In this hypothetical case they might add: "However, due to sampling error of five percentage points [this is also referred to as a confidence limit] the election appears to be a dead heat and too close to predict a winner." Similar statements are made during actual elections when newscasters might report "With 10% of the precincts reporting, we are predicting candidate C to be the winner over candidate D by a margin of 60% to 39%." What do these figures represent? How do they arrive at these figures? And with 52% clearly being more than 47%, and a large enough margin to win most elections, why is the election still reported as a dead heat?

Answering each of these questions in turn, the figures of 52% and 47% represent proportions of a sample of registered voters in terms of which candidate they plan to vote for in the upcoming election. In other words, referring all the way back to Chapter 1, these proportions are inferential statistics calculated from a sample that was drawn from a larger population. [As an aside, if these figures (52% and 47%) were for an upcoming presidential election and we wanted to predict the winner, the population our sample would be drawn from comprises all registered voters in the United States. Conversely, if the figures were for the elected position of mayor of the town or city you reside in, the population would be all registered voters in your town or city.]

How these figures are actually calculated and why they point to a dead heat is what the rest of this chapter is devoted to explaining. At this point, however, there are some simple observations we can make. With the above values of 52% and 47% with corresponding margins of sampling error equal to 5%, we are actually being told that the true percentage of people who supports candidate A is somewhere between 47% and 57% (52% ± 5%) of the voters while the proportion that supports candidate B is somewhere between 42% and 52% (47% ± 5%). Obviously, since there is considerable overlap between the actual proportions of people that may support candidate A versus candidate B, we cannot state conclusively that either candidate appears to have a clear-cut lead.

To answer the above questions more completely, this chapter discusses how to calculate the statistical concepts of the standard error of the mean, confidence limits and levels (using the previously discussed Z distribution and a newly-introduced t distribution), and alpha levels; altogether, these terms equal a confidence interval. We also explore a statistical technique that tells us the sample size required for a stated margin of error. While all these new terms may seem somewhat overwhelming, in application they are, as usual, quite easy to calculate and understand. Quite simply, what confidence intervals do is estimate population parameters. While an array of different confidence intervals exists to estimate nearly every conceivable population parameter, to keep things simple we discuss only estimates of population means in this chapter.

Samples

Before any actual calculations for confidence intervals are undertaken, we should review briefly what samples and statistics enable us to do. To assist in this discussion, we have recreated below the figure that initially appeared in Chapter 1 (Figure 8.1). The importance of this figure to this chapter's material is that it points out the two things that samples and corresponding statistics do in reference to population parameters: estimate and hypothesis test. The former, estimates of population parameters, is what this chapter is all about, while the latter, hypothesis testing, is what Chapters 9 through 12 are largely about.

To this point we have primarily been concerned with two different types of sample statistics: means and standard deviations. Until now, however, we have had no way to assess how accurate these and other statistics are in terms of the population parameters they estimate. That is, while sample means, standard deviations, and other descriptive statistics are the best point estimates we have for each corresponding parameter, these figures by themselves tell us nothing about how accurate they are. Accuracy, in this context, means how much the sample statistic potentially deviates from the parameter it is estimating.

This is exactly what confidence intervals do; they enable us to determine the accu-

FIGURE 8.1 Population Parameters/Sample Statistics

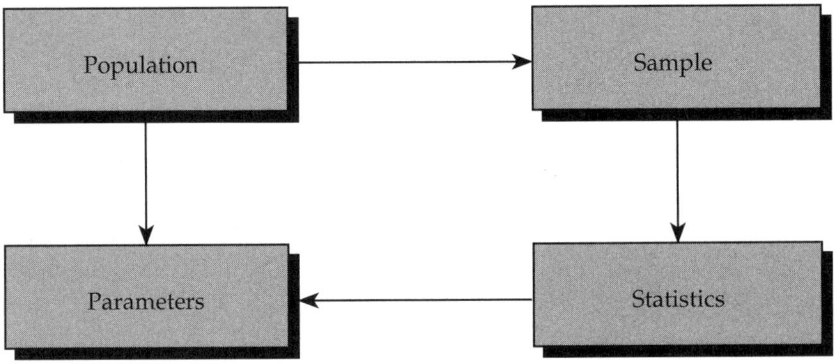

1. Estimate Population Parameters.
2. Hypothesis Testing.

racy of our initial estimates. To accomplish this, information from the population and the sample (or, more typically, just from the sample) is used to calculate the given estimate's accuracy. Moreover, and indirectly based upon the material discussed in the previous two chapters, confidence intervals also make estimates of accuracy in terms of probability values. In sum, *confidence intervals* are probability estimates of the true parameter value in terms of its occurrence between constructed boundaries. This will become clearer as we proceed.

Confidence Intervals Using a Population Standard Deviation

Below is the formula used to calculate a confidence interval when the population standard deviation is known. Given this information, what the formula estimates is an unknown population mean.

$$CI = \bar{x} \pm (Z_{\alpha/2})\left(\frac{\sigma}{\sqrt{n}}\right)$$ Confidence Interval When σ is Known

Before we explore the formula in its complete form, it is much easier to introduce individually and explain each of its four sub-parts: the *standard error of the mean*, the *confidence level*, the *alpha level*, and the *confidence limits*. Once each of these components is explained, there is only one more step required to construct a confidence interval.

Cartoon 8.1 serves as a backdrop for the discussion that follows in this section. This cartoon is representative of a measure that we have administered in many of our sociology courses (including statistics) called the BEM scale. The BEM scale measures something called psychological androgyny. To this end, the scale is comprised of 60 questions that assess each respondent's stereotypical attributes of instrumentalness (construed as traditional masculine characteristics—Ward Cleaver) and expressiveness (construed as traditional feminine characteristics—June Cleaver). Of the 60 questions, 20 measure instrumentalness, 20 measure expressiveness, and 20 are filter items that don't measure anything per se. Further, the 60 questions ask the respondents to answer how applicable each one-word item is to them on a seven-point scale. After the scale is completed, the 20 masculine scores and the 20 feminine scores are separately added together, and then each divided by 20—this gives two average scores: one for masculinity and one for femininity.

CARTOON 8.1

(As an aside, respondents whose average masculine and feminine scores are within one whole point of each other are considered psychologically androgynous. This means that, psychologically, they have nearly equal amounts of traditional male and female personality characteristics. Many contemporary social commentators view psychological androgyny as a positive personality trait.)

We should also note that while the BEM scale is an ordinal level measurement, the confidence interval statistical technique was initially created with the assumption that only interval/ratio level data should be used. The reason for this assumption is beyond the scope of this text. Needless to say, as with our example, this assumption is regularly violated; that is, it is rather commonplace for social scientists to construct confidence intervals using ordinal level data. Since violating this assumption potentially draws into question our findings, however, we should always make note of circumstances where lower levels of measurement have been used.

Before proceeding, we should quickly review what is found in the confidence-interval formula. As noted about the original formula, we are assuming that the population standard deviation value is already known. Although it is actually fairly uncommon to know the population standard deviation when calculating a confidence interval, there are instances, such as in industry, where this value may be known. Further, as demonstrated in the next section, using a population standard deviation versus a sample standard deviation entails the use of different set of standardized scores (a corresponding Z distribution versus the to-be-introduced t distribution) to calculate a confidence interval.

With the above provisions in mind, we are finally ready to apply an actual data set to the abstract discussion we have thus far undertaken. The first author administered the BEM scale in a course he taught and found the 36 students that attended class that day had an average expressive score of 4.2 and an average instrumental score of 4.8 (once again, on a scale of 1 to 7). Let's say additionally we know that the standard deviation values for *all* the students at the university he teaches are 1.5 for the expressive scores and 1.8 for the instrumental scores. Thus, in this case, we are treating our 36 students as a nonrandom sample and *all* students at the university as the *population.* (Ideally, we would have generated a random sample from all of the university students. Only then can we assume that our sample is truly representative.)

Utilizing the information from this data set, we will first construct a confidence interval for the expressive scores to estimate what the actual expressive average (population mean) is for all students at the university. To this end, one of the first things we can do is calculate the *standard error of the mean.* To determine this value, we take the population standard deviation and divide it by the square root of the sample size. This sub-part of the overall confidence interval formula is mathematically represented below.

$$\sigma/\sqrt{n}$$

Formula for the Standard Error of the Mean

Without miring ourselves in too technical a discussion, there are still some important observations that we can make about what the standard error of the mean represents. Once again, a standard deviation represents the amount of standardized variance found in a given set of scores. Since we divide the standard deviation by the square root of the sample size (n), the size of the final answer for this operation is directly determined by both the amount of standardized variance and the sample size. To state this in slightly different terms, the larger the standard deviation and/or the smaller the sample, the larger the standard error of the mean will be. Conversely, the smaller the standard deviation and/or the larger the sample, the smaller the standard error of the mean will be. As we will see, since the size of a confidence interval is a measure of the accuracy of our estimate of the population mean, a smaller standard error of the mean value is always desired. In other words, using a smaller standard error value results in a more accurate estimate.

Below we have placed the appropriate BEM expressive values into the standard error of the mean formula ($n = 36; \sigma = 1.5$) to obtain the answer of .25.

$$\frac{\sigma}{\sqrt{n}} = \frac{1.5}{\sqrt{36}} = \frac{1.5}{6} = \boxed{.25^*}$$

*Standard Error of the Mean For the BEM Expressive Scores

The next sub-part that we need to calculate and define is the *confidence level.* This is represented by the portion of the confidence interval formula that has the notation of a Z with a new subscript ($\alpha/2$). What it represents is an actual Z value. To keep things simple, for our purposes this is always one of two Z-score values: 1.96 and 2.58. That is, when we undertake subsequent calculations for confidence intervals where the population standard deviation is known, one of these two values is always used.

Although the values of 1.96 and 2.58 are now treated as givens, it is still helpful to discuss briefly how they are derived. These values reflect two different confidence levels. Taking the Z value of 1.96 first and tracing its calculations, we must define what it represents: a confidence level, which in this case is 95%. A confidence level of 95% informs us that our final answer, the confidence interval itself, will contain the true population mean that it is estimating 95 times out of 100. This is more clearly demonstrated shortly.

A 95% confidence level also tells us that we have an error level of 5%; that is, 5% of the time our estimate is wrong—in error. The error level is more correctly referred to as an *alpha level* and is represented by the symbol α, the subscript found next to the Z. To determine an alpha level, we simply take the value of one and subtract the given confidence level from it. In the present example, this is $1 - .95 = .05$ or a 5% alpha level. Next, as the subscript instructs us, we divide the alpha level by 2; for this example, $.05/2 = .025$. Referring back to Appendix 1 and its corresponding Z scores, we look in the Small-Part column for the probability value of .025 and find that the attendant Z value is 1.96.

Reapplying these steps to the Z value of 2.58, we first must note that this value represents a confidence level of 99%. This tells us that the corresponding alpha level is 1%, derived by taking $1-.99 = .01$. Next, we take the alpha level of .01, and divide it by 2, to give us the value of .005. Once again, we look this value up in the Small-Part column found in Appendix 1, and find that it falls between two Z scores: 2.57 and 2.58. Since many statisticians often use the larger value of 2.58 to simplify subsequent calculations, we also use this value.

Now that we have determined how the values of 1.96 and 2.58 are derived, before we proceed any further a few more important observations are warranted. The values of 95% and 99%, as previously noted, are the confidence levels most commonly used in the social sciences. For this reason, and to simplify our discussion, only these confidence levels are used for Z scores. In other words, we no longer have to refer to Appendix 1 or do any further calculations for this component of the formula. It is important, however, to note that numerous other confidence levels, such as 98% or 99.9%, can also be used (the next type of confidence interval discussed in this chapter will consider these levels). Finally, the confidence levels of 95% and 99% and their corresponding alpha levels of .05 and .01, in reference to Z scores, represent the same thing and are sometimes used interchangeably in subsequent discussions.

We are now ready to calculate the *confidence limits*: the final sub-part of the confidence interval formula. Confidence limits are sometimes also called the "margin of error," and in the next section they are referred to as such. To calculate the confidence limits, the appropriate Z value (1.96 or 2.58) is multiplied by the standard error of the mean. Since the confidence level is usually arbitrarily set (usually determined by how accurate an estimate of the population mean the given researcher wants), the calculations for both the 95% level and the 99% level are completed below.

<u>Confidence Limit</u>

$(Z_{\alpha/2})\left(\frac{\sigma}{\sqrt{n}}\right)$ for

95% level = (1.96) (.25) = $\boxed{.490}$

99% level = (2.58) (.25) = $\boxed{.645}$

All that's left to do to get our final answer, the *confidence interval,* is to add and subtract (±) the confidence limits from the sample mean; for this case, recall that our sample of 36 students had an expressive average of 4.2. Found below is the original confidence interval formula, all of the original values for each confidence interval, the calculations for both confidence levels (.05 and .01), and the two resultant confidence intervals.

Confidence Intervals for the Alpha Levels of .05 and .01

$$CI = \bar{x} \pm (Z_{\alpha/2})(\sigma/\sqrt{n})$$

For Alpha Level .05

$$4.2 \pm (1.96)(1.5/\sqrt{36}) = 4.2 \pm .49 = \boxed{(3.71 < \mu < 4.69), .05}$$

For Alpha Level .01

$$4.2 \pm (2.58)(1.5/\sqrt{36}) = 4.2 \pm .645 = \boxed{(3.555 < \mu < 4.845), .01}$$

The presentation format of these confidence intervals and why there is a difference between the two warrants some further discussion. Since the given confidence interval is an estimate of the true population mean, the population mean is placed in the middle of the constructed interval. Directly to the left and right of the population mean (μ; "mu") are two symbols that signify that the true mean is greater than the lowest part of the interval estimate (referred to as a *lower limit*) and less than the highest part (referred to as an *upper limit*). (As a somewhat ridiculous aside, some of our students have inferred the signs of greater and less than to be the equivalent to the fast forward and reverse buttons on their VCR's.) Directly to the right of the given interval is the alpha level used in its calculation.

To put all of this into English, the first confidence interval informs us that the population mean for expressive scores (June Cleaverness) at the university is somewhere between the values of 3.71 and 4.69. Further, with the alpha level set at .05 (confidence level 95%) we assume that this estimate is correct 95 times out of 100; 95% of the time the true population mean will fall between 3.71 and 4.69. Conversely, we also assume that the estimate is in error 5% of the time; we expect the true population mean to fall outside of the above confidence interval 5 times out 100. In total, we can state that we are 95% confident that the population mean for expressiveness at the university falls between the values of 3.71 and 4.69. Alternatively, for the second confidence interval we are 99% confident that the population mean for expressiveness falls between the values of 3.555 and 4.845.

Most of you have probably noticed that as our confidence level goes up (in this example, from 95% to 99%), the corresponding width of the confidence interval also increases. The reason for this (beyond the obvious larger Z value used in the calculation of the confidence interval) is that as we become more confident that the true population mean is found within our estimate, the size of the interval must increase. A helpful way to demonstrate this relationship is to superimpose both of the confidence

FIGURE 8.2 Z Distribution

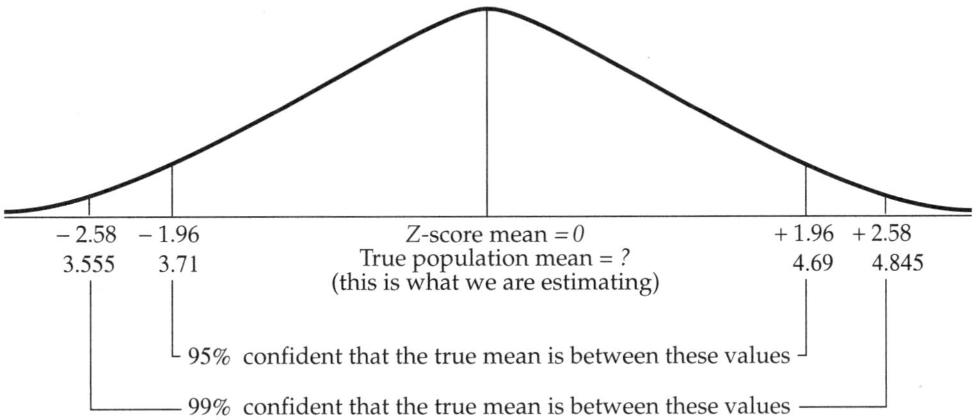

intervals we have constructed onto the normal (Z score) distribution. This is done in Figure 8.2.

In a simplistic sense, by applying the constructed intervals onto the normal distribution we theoretically demonstrate that our constructed interval represents 95% or 99% of all estimated scores. Thus, correspondingly, this is how confident we are that the population mean we are estimating falls between the given interval's upper and lower limits. After all, any given mean is calculated from a distribution of actual scores.

For review purposes, let's construct the two confidence intervals for the instrumental scores (Ward Cleaverness). The mean instrumental score for the 36 students, once again, was 4.8 with an already known population standard deviation of 1.8. Instead of going through each of the sub-parts to calculate this confidence interval, the reproduced formula and the calculations for each alpha level (.05 and .01) are simply presented below. If any of the values or calculations are problematic to you, please refer to the step-by-step discussion just undertaken for the expressive scores.

Confidence Intervals for the Alpha Levels of .05 and .01

$$CI = \bar{x} \pm (Z_{\alpha/2})\,(\sigma/\sqrt{n})$$

For Alpha Level .05

$$4.2 \pm (1.96)\,\underset{(.3)}{(1.8/\sqrt{36})} = 4.8 \pm .588 = \boxed{(4.212 < \mu < 5.388), .05}$$

For Alpha Level .01

$$4.2 \pm (2.58)\,\underset{(.3)}{(1.8/\sqrt{36})} = 4.8 \pm .774 = \boxed{(4.026 < \mu < 5.574), .01}$$

Before proceeding, we need to make note of two of the most obvious differences between the above confidence intervals and those previously constructed. First, since the instrumental mean is .6 greater than the expressive mean, the constructed intervals cover a different part of the 1–7 score range of the BEM scale. Second, since the standard deviation for the instrumental scores (1.8) is larger than that for the expressive scores (1.5), the widths of the resultant confidence intervals are larger: .98 at .05 and 1.29 at .01 for the expressive scores versus 1.168 at .05 and 1.548 at .01 for the instrumental scores. This demonstrates the importance of the already noted size of the standard deviation; as it increases the width of the confidence interval also increases. Of course, once again, one way a researcher can compensate for this is to increase the sample size, which always decreases the width of the confidence interval. This relationship is more clearly demonstrated in the next section.

Summarizing the above confidence intervals, at the .05 alpha level we can estimate that 95 times out of 100 the population mean of instrumental scores for all university students will fall between 4.212 and 5.388; 5 times out of 100 it will fall outside. At the .01 alpha level we can estimate that 99 times out of 100 the population mean will fall between 4.026 and 5.574; 1 time out of 100 it will fall outside.

Sample Size Required for a Given Margin of Error

When the population standard deviation is known, as with the above confidence intervals, this information can help researchers and pollsters estimate what size sample is required for a predetermined margin of error. As already noted, the *margin of error* is the same thing as a *confidence limit.* Mathematically, when the population standard deviation and confidence level are both known, we can predetermine the sample size required for a given confidence interval. The formula to do this is derived from the original confidence-interval formula using the same simple algebraic logic that tells us that 1=1/2 thus 2=1/1 or 2=1/4 thus 4=1/2. By algebraically transforming the original confidence interval formula (e = the margin of error) we find

$$\text{If } e = (Z_{\alpha/2})\left(\frac{\sigma}{\sqrt{n}}\right), \quad \text{then } n = \left(\frac{(Z_{\alpha/2})\,(\sigma)}{e}\right)^2$$

One very practical application of this formula that corporations often use is to test products—quality control. For instance, light bulb companies often make the claim that a given bulb lasts 400 hours or batteries for portable radios and mechanical rabbits will last 40 hours. Taking the latter example of batteries and mechanical rabbits, let's say we want to know what size sample of batteries is required (number of batteries to be tested) to have a 2-hour margin of error. From previous tests we know the population standard deviation is 4 hours. All we have to do to complete the formula is set the alpha level; for the first calculation it is set at .05. The original formula, all relevant values ($Z = 1.96$, $\sigma = 4$, and $e = 2$), and all calculations are found below.

$$n = \left(\frac{(Z_{\alpha/2})\,(\sigma)}{e}\right)^2 = \left(\frac{(1.96)\,(4)}{2}\right)^2 = \left(\frac{7.84}{2}\right)^{2*} = (3.92)^2 = \boxed{16^{**}}$$

* We must also note that students using this formula often forget to square the final answer; make sure you don't!

** One immediate point of clarification:while 3.92 squared equals 15.3664, since we are dealing with whole sampling units required for a given margin of error, always round up to the next whole unit. After all, it would be impossible to have a sample of 15.3664 batteries.

The final answer tells us that we need a sample of 16 batteries to have a confidence limit of 2 hours. Further, we are 95% confident in this estimate of the sample size required. To validate this conclusion, we can take the predicted sample size required (15.3664 for the actual calculation) along with the other given values and plug them into the original confidence interval formula to find that

$$e = (Z_{\alpha/2})\,(\sigma/\sqrt{n}) = (1.96)\,(4/\sqrt{15.3664}) = \boxed{2}$$

Alternatively, let's say that we want the margin of error for our batteries to be within one half hour (.5). Using the same alpha level of .05 and population standard deviation as in the previous example (4), but different margin of error (e = .5) in this case, we have completed the appropriate calculations below.

$$n = \left(\frac{(Z_{\alpha/2})\,(\sigma)}{e}\right)^2 = \left(\frac{(1.96)\,(4)}{.5}\right)^2 = \left(\frac{7.84}{.5}\right)^2 = (15.68)^2 = \boxed{246}$$

Thus, to decrease the margin of error by 1.5 hours (from 2 hours to .5 hours), we must drastically increase the sample size from 16 to 246 batteries. Referring back to the original confidence-interval formula, this relationship clearly demonstrates the importance of sample size in determining the width of a confidence interval.

As already noted, the alpha level has the same effect on the sample size. For instance, in our first example above, with the alpha level at .01, we need a sample of 27 instead of 16, while the sample size required for the second example would increase from 246 to 427. All of this has strong implications for pollsters and industry when determining how large a sample to draw for their research. That is, it costs money to increase a given sample size, and to be just a little more accurate or confident in one's estimate is often not worth it.

Confidence Intervals Using a Sample Standard Deviation

Thus far, our discussion of confidence intervals has assumed that we already know what the population standard deviation value is; however, as already noted, we seldom know this value. Like a sample mean, the best estimate of a population standard deviation is the sample standard deviation. Below is the formula used to calculate a confidence interval using a sample standard deviation.

$$CI = \bar{x} \pm (t_{\alpha/2})\,(s/\sqrt{n}) \qquad \text{Confidence Interval Using } s$$

Although this new formula appears very similar to the previous one used, there are also some important differences. To begin with the obvious, instead of using a population standard deviation, we now use a sample standard deviation value. Moreover, while the use of a sample standard deviation value does not change the format of the actual calculations, it does require that we now use a t instead of a Z value. As such, before we can actually construct a confidence interval with this new formula, we must explain its accompanying distribution and how to use it.

Box 9

> The t distribution is actually called the Student's t distribution. The creator of this distribution, William Gosset, was an employee of Guinness Brewery in Dublin, Ireland. He developed this distribution as a technique to deal with small samples of barley and hops and other conditions of production to ensure the quality of his brewery's product. The reason that this distribution does not bear his name is that one of the conditions of his employment was that he could not personally take credit for any of his publications; he simply signed all his works "Student." This is why this distribution is called Student's t.

The *t distribution* is found in Appendix 2 (please refer to it for the following discussion). The two pages of this appendix have a format different than the one used for the Z distribution. Instead of column headings for what part of the distribution we are concerned with, the column headings for the t distribution have alpha levels. One set of column headings is for *one-tailed values* whereas the second set of column headings is for *two-tailed values.* Since the easiest way to construct this type of confidence interval is with two-tailed values, only the two-tailed column headings and their corresponding values are used in this chapter.

Next, on the left side of the table are found what are called degrees of freedom, signified by "df." Degrees of freedom are determined by subtracting the value of one from the sample size; $n-1$. The reason that degrees of freedom are determined using an $n-1$ is that we are estimating a population parameter; this is the same reason the denominator in the standard deviation formula is $n-1$. In other words, since we are estimating a population mean, the absolute minimum number of values (observations) that would potentially have to change for our estimate to equal the mean is one. Thus, fixing one degree of freedom leaves us with an estimate that takes into account potential sampling error.

When these two pieces of information are combined, alpha level and df, we can determine any t-critical value. Before we undertake the complete confidence-interval formula, let's first practice determining some t-critical values. Given a sample of 16 and an alpha level of .05, what is the corresponding t-critical value? First we determine the degrees of freedom by simply subtracting one from the sample size; $16 - 1 = 15$ df. Next we look down the row headings to 15 df, and then proceed to the column heading of alpha .05, to ascertain that the t-critical value for this example is 2.131.

For the next example, what is the corresponding t-critical value for a sample of 10 with an alpha level of .001? (As an aside, a .001 alpha level is the same thing as a confidence level of 99.9%; that is, the resultant confidence interval, if it were completed for this example, would contain the true mean value an estimated 999 times out of 1000.) To answer the question, we first determine the degrees of freedom (10 − 1 = 9 df), find this row heading combined with the column heading of .001, and find the *t*-critical value of 4.781.

Finally, let's say we have a sample of 37 and want to know what corresponding *t*-critical value is at the .02 alpha level. The .02 column heading is easy to locate, but what do we do with 36 df (37 − 1 = 36 df) when this row heading is not listed? In situations such as this where the df value falls between two listed values), we ask that you always take the next lowest value on the table, in this case 30 df. This gives us a more conservative estimate and saves us from doing an additional step called extrapolation. Thus, with 36 df (30 df) at the .02 alpha level the corresponding *t* value is 2.457.

Box 10

Although the *t* distribution is very similar to the *Z* distribution (since it was created specifically to deal with smaller sets of numbers, samples, especially those under 30), there are also some significant differences. To begin with, as illustrated in Figure 8.3, the *t* distribution is much flatter, with thicker tail regions than the *Z*. The reason for this is that as a sample size decreases the corresponding potential for sampling error increases. Stating this in slightly different terms, as the sample size decreases, the scores tend to become more evenly disbursed throughout the distribution. This is in direct contrast to the *Z* distribution, where most of the scores tend to occur right around the mean.

To further demonstrate this difference between distributions, with the *Z* distribution at plus or minus two standard deviations (1.96 to be exact), we expect to find 95% of all the scores in this region. However, with a sample of 6 (5 df) it takes over two and one-half standard deviations (2.571) to make this same conclusion. Both of these examples are also illustrated in Figure 8.3.

On the other hand, as the sample size increases, the shape of the *t* distribution becomes more and more like that of the *Z* distribution. Referring back to Appendix 2, if one looks at the row heading of 120 df under the .05 column heading, the critical value is 1.98, almost the same as that of *Z* distribution at this level. Moreover, just below this value is the row heading of infinite (∞) and the expected 1.96.

We are now ready to construct a confidence interval using this new distribution. To assist in this discussion, we offer Cartoon 8.2, "A Student Lives!" More specifically, our current research interests are the number of "thinking responses" or inquiries students make per semester (or quarter) in a statistics class taught with the text that you are presently using. We randomly sample 16 students currently enrolled in statistics classes at a large university that are using this text to determine the average number of

FIGURE 8.3 *Z* and *t* Distributions

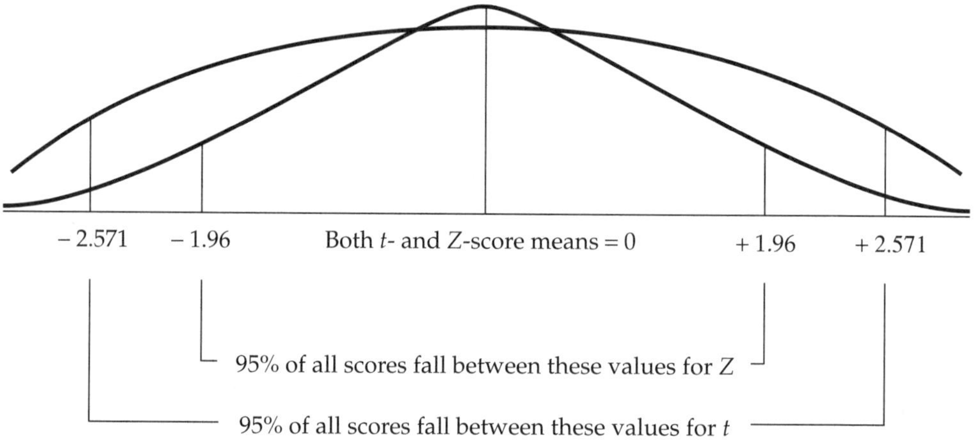

"thinking responses" they make in these types of classes. They are then watched for a semester's period. We find that the average ($\bar{x}$) of "thinking responses" is 30 with a standard deviation (s) of 5 such responses for our sampled 16 students.

With the alpha level set at .05 and .001, we use this information to construct two confidence intervals to estimate what the population mean is. The population, in this case, is all students attending statistics classes at this given university that are using this text.

We use the following four basic steps to construct this type of confidence interval:

DOONESBURY by Garry Trudeau CARTOON 8.2

(1) Derive the standard error of the mean, (2) determine the t-critical value, (3) multiply these two values together, to give us our confidence limits, and, finally, (4) add and subtract the confidence limits from the sample mean to give us the confidence interval. The formula to do this and all these steps for both confidence intervals (.05 and .001) are completed below; however, since this is our first example of this type of confidence interval, we still briefly outline each step for the .05 interval.

Once again, to determine the standard error of the mean we take the sample standard deviation (5) and divide it by the square root of n ($\sqrt{16} = 4$); 5 divided by 4 equals 1.25. Next, refer to Appendix 2 to find that the t critical value is 2.131; column heading of .05 with 15 degrees of freedom ($n - 1$; $16 - 1 = 15$ df). We multiply these two values together (the t critical by the standard error of the mean) to derive the confidence limit of 2.6637. Finally, we just add and subtract this value from our sample mean (30 thinking responses) to obtain the confidence interval of ($27.3363 < \mu < 32.6637$), .05.

Confidence Intervals for Alpha Levels Set at .05 and .001

$$CI = \bar{x} \pm (t_{\alpha/2})\,(s/\sqrt{n})$$ Confidence Interval Using s

For Alpha Level .05

$$30 \pm (2.131)\,\underset{(1.25)}{(5/\sqrt{16})} = 30 \pm 2.6637 = \boxed{(27.3363 < \mu < 32.6637), .05}$$

For Alpha Level .01

$$30 \pm (4.073)\,\underset{(1.25)}{(5/\sqrt{16})} = 30 \pm 5.0912 = \boxed{(24.9088 < \mu < 35.0912), .001}$$

Summarizing these confidence intervals, for the first one with the alpha level set at .05, we are 95% confident that the true sample mean occurs between the values of 27.3363 and 32.6637. In other words, we are estimating that the average number of "thinking responses" made by statistics students using this text at our hypothetical university (the population) is found between these values 95 times out of 100; 5 times out of 100 the population mean falls outside of this interval.

The second confidence interval, alpha level of .001, is much larger. The reason for this, once again, is that we are much more confident that the true population mean is found within our constructed interval. Stated in the context of the resultant answer, 999 times out of 1000 the true population mean is expected to occur between the values of 24.9088 and 35.0912 and only 1 time out of 1000 is it expected to fall outside of this interval. Thus, we can be quite confident that most students using this text at this university will make somewhere between 25 and 35 "thinking responses" per semester.

Chapter Summary and Conclusions

This chapter began with a discussion of inferential statistics and what they enable us to do: estimate population parameters and hypothesis test. Dealing exclusively with the former, two techniques to estimate population means were outlined. The first technique was a confidence interval when the population standard deviation is known. We noted that since the population standard deviation is typically not known, this type of confidence interval is not widely used. Next we outlined an algebraic variant of this first confidence-interval formula that enabled us to determine the sample size required for a given margin of error. The second, more widely used technique that we next explored utilized a sample standard deviation and a newly introduced t distribution to construct confidence intervals. We also made note that there are other techniques of estimation for proportions and other population parameters. The next chapter builds upon the confidence intervals presented in this chapter and uses them to investigate the second thing sample statistics enable us to do: hypothesis test.

KEY TERMS TO REMEMBER

Samples
Inferential Statistics
Populations
Parameters
Alpha Levels
Confidence Intervals
Standard Error of the Mean
Margin of Error
Confidence Level
Confidence Limits
Z Distribution
t Distribution

PRACTICE EXERCISES

1. Using Cartoon 8.2 again as a backdrop, let's say that we are interested in the number of "thinking responses" made in a statistics class that does not use this text. We find the following:
   Mean ($\bar{x}$) = 22 Thinking Responses
   Population Standard Deviation (σ) = 4.5 Thinking Responses
   $n = 25$
   a. Calculate the standard error of the mean.
   b. With the confidence level set at 95%, what are the confidence limits for this mean?
   c. Construct a confidence interval using the confidence limits.
2. Let's say that we are given the following information on the average life of a 100-watt light bulb:
   Mean ($\bar{x}$) = 400 hours
   Population Standard Deviation (σ) = 36 hours
   $n = 144$
   a. Calculate the standard error of the mean.
   b. With the confidence level set at 99%, what are the confidence limits for this mean?

c. Construct a confidence interval using the confidence limits.

3. Let's say that we are given the following information on the average amount of time spent daily in lines (such as at the grocery store, the bank, or any of the numerous mazes found in your college's or university's administration building—all great places to spend time waiting to give other people your money):
Mean ($\bar{x}$) = 43 minutes
Standard Deviation (s) = 12 minutes
$n = 36$
   a. Calculate the standard error of the mean.
   b. With a confidence level set at 95% and 99%, what are the different confidence limits (2) for this mean?
   c. Construct *two* confidence intervals using the two different confidence limits.
   d. Explain why there is a difference between the two confidence intervals' widths.
4. Using the information in question 3:
   a. Repeat the questions a, b, and c (using just the 95% confidence level) with an n of 120.
   b. Why does increasing the sample size have such a drastic impact on the width of the resultant interval?
5. For females aged 40–49 in the United States, the average weight is 142 pounds, with a standard deviation of 27 pounds.
   a. How large a sample would we need to estimate the mean with a margin of error of 2 pounds? (alpha level .05 and .01)
   b. One-half pound? (alpha level .05 and .01)
   c. Which sample size is more economically feasible?

Chapter 9

Hypothesis Testing between Two Sets of Observations

Having explored how confidence intervals are constructed and what they represent in the last chapter, we are now ready to take these ideas one step further and show how they are also used to test hypotheses. (Recall from Chapter 1 that estimating population parameters and hypothesis testing are the two things that inferential statistics enable us to do.) While the construction of confidence intervals is important to social scientists who undertake quantitative research, perhaps an even more important function that they enable us to do is to test hypotheses. Quite simply, *hypothesis testing* is what almost all quantitative studies and journals ultimately report; these are the all-important findings. As we will see, it is through the process of hypothesis testing that the phrase "statistically significant" is derived.

Specifically, this chapter first explores the underlying assumptions of hypothesis testing and statistical significance. These ideas are then applied to four types of statistical tests (ways of hypothesis testing); a *Z test*, *a dependent-sample t test*, an *independent-samples t test*, and a *matched t test*. Because the rest of the text is largely devoted to hypothesis testing, these four initial tests are viewed as starting points for understanding statistical testing.

Hypothesis Testing

All scientists are ultimately interested in explaining how and why certain things occur the way they do. For social scientists, this interest is obviously in terms explaining social phenomena. Social phenomena are a nearly inexhaustible array of every imaginable attitude, behavior, and social characteristic that one can think of. What social scientists attempt to do is establish relationships of association (two variables tend to occur together) and causality (one variable is seen as bringing about a change in another variable) between different social phenomena. Another way of viewing social phenomena is in terms of a measurable variable. For example, a measurable attitude is whether one supports the legalization of marijuana, a measurable behavior is if one smokes marijuana, and a measurable social characteristic is one's age. One possible relationship among these three variables might be that those who support marijuana legalization are more likely to smoke it than those who don't support legalization. Moreover, we might expect that younger people are more likely to support marijuana legalization and to smoke it than older people. If this in fact was found to be true, we would conclude that age and marijuana use are both determinants—seen as causes—of attitudes about legalization.

Box 11

Physical scientists' research interests, in many ways, are far different than social scientists. Physical scientists are interested in explaining things such as why chemical and biological reactions occur the way they do. Social scientists, on the other hand, are interested in explaining social phenomena. This differing emphasis has strong implications in terms of how research is undertaken. For physical scientists, most of the research they undertake is in the artificial setting of the laboratory. Such a setting enables the scientist to control for factors (variables) that may influence research outcomes (e.g., how much of a given element or chemical is used).

Although a limited amount of social research is done in laboratory settings, such as behavioral research undertaken by psychologists, most of it is done outside of this setting in the real world where we all reside. As such, whereas physical scientists can often explain the exact effect one single variable has upon another single variable, this is simply not possible with social research. Since social scientists deal with human behavior, they are faced with research situations where numerous variables can be used to explain one variable. Moreover, some of the variables are extraneous and just associated with, but not an actual cause of, a given variable. Further, even the most carefully designed social research project seldom, if ever, takes into account all of the variables that are actually the cause of another variable. Even in laboratories, where human subjects are involved, these same problems are prevalent.

As a result, there are many more complexities that a social scientist must deal with when she or he undertakes research. This also poses problems with statistical testing. Quite simply, while this chapter deals exclusively with *bivariate* relationships (two variables, where one is seen as potentially affecting the other), almost all social relationships are *multivariate*. That is, no one variable is the exclusive cause of change in another. More typically, there are always numerous variables that explain a given outcome, with many of them often not even being measured.

A more formal way of viewing variable relationships of association and causality is in terms of a stated hypothesis. A stated hypothesis actually comes in two basic forms: the *null hypothesis* and the *research hypothesis* (also called the alternative hypothesis). The null hypothesis simply means *no difference*, one variable is not associated with another. Conversely, the research/alternative hypothesis simply means there is a relationship of association. For the purposes of this chapter, one way that we can determine if one variable is associated with or brings about a change in another variable is by comparing group means (averages) associated with a given bivariate relationship. This is actually what all the aforementioned statistical tests in this chapter do. What these tests allow us to do is to determine, statistically speaking, if there is a difference between two group means. One generic way this is presented is found below; the actual means tested for a difference varies from test to test.

Null Hypothesis	$H_0 : \bar{x} = \bar{x}$
Alternative or Research Hypothesis	$H_A : \bar{x} \neq \bar{x}$

As the above symbolically demonstrates, the null hypothesis tells us that there is *no* difference between the means; they are statistically equal. On the other hand, the research hypothesis implies that there is a difference between two means. This is mathematically represented by the symbol of "not equal" between the two means.

Before turning to any calculations, let's apply what we have discussed thus far to Cartoon 9.1 (we use this same information when we undertake an actual test). Let's say our current research interests are how intelligent—or dimwitted—different breeds of cows are. Reflecting on the cartoon, we decide to measure intelligence in terms of the number of times per day a cow touches an electric fence. That is, the fewer the number of times a cow runs into the electric fence, the more intelligent it is perceived as being. Let's say we know that the daily average that all cows (all breeds) in the United States touch an electric fence is 10, with a standard deviation of 1.5 times. We want to know if the Holsteins are more or less intelligent than all breeds of cows combined. We sample 25 Holstein cows and find that on the average they touch an electric fence 7 times a day.

Stating this in terms of hypothesis testing, the null hypothesis holds that, statistically speaking, there is no difference between the average number of times a Holstein runs into the electric fence and the national average for all cows. We need to note that just because a numerical difference may exist between two means does not necessarily mean that there is a statistical difference. As we will see, nonsignificant differences are due to potential sampling error. The research hypothesis, on the other hand, holds that there is a statistical difference between the means. This is symbolically represented below. [μ represents the national average (population) while the $\bar{x}$ represents the Holstein average (sample).]

Null Hypothesis	$H_0 : \mu = \bar{x}$
Alternative or Research Hypothesis	$H_A : \mu \neq \bar{x}$

Cow poetry CARTOON 9.1

There are obviously two possible outcomes in reference to these two statements; there *is* or *is not* a statistical difference between the two group means. These two outcomes can be stated in more scientific terms. In the first case, if there is a statistical difference between the two means, we *reject* the null hypothesis and *accept* the alternative (research) hypothesis. Conversely, if there is not a statistical difference between the two means, we *accept* the null hypothesis and *reject* the alternative hypothesis.

We reject or accept the null or research hypotheses, like confidence intervals, using an alpha level. That is, dependent upon the given alpha level, we can state how confident we are that the null hypothesis has been correctly rejected. The alpha level also tells us the probability that we have rejected the null hypothesis when in fact it is true. If this is done (i.e., the null hypothesis is rejected when it's in fact true), we have committed what is referred to as a *Type I Error* (this term is discussed in detail in a moment).

Putting this into English, let's say the above bovine (cow) example has a preset alpha level of .05 and we find that the difference between the number of times per day a Holstein touches an electric fence (7) and the national average (10) is statistically significant. As such, Holsteins are far less likely, on the average, to touch the "damn" electric fence, and thus (because of this difference) we infer that they are more intelligent. Moreover, we are 95% confident that we have correctly rejected the null hypothesis. On the other hand, we also recognize that there is a 5% chance that we have rejected the null hypothesis when in fact it is true: a Type I Error. When the first actual hypothesis test is undertaken, in a moment, this becomes clearer.

Although beyond the scope of this text in terms of how it is numerically calculated, it is nevertheless important to discuss a second type of statistical error that can occur when hypothesis testing: a *Type II Error*. This type of error occurs when we do not reject the null hypothesis when in fact we should have. In other words, we fail to recognize a statistically significant difference between two means when in fact there is one. Since a great deal of research undertaken hopes to find a difference (some researchers' funding may even be contingent upon finding a difference), this can be very problematic. Both the Type II Error and previously discussed Type I Error outcomes along with the relationships of the correct rejection and acceptance of the null and research hypotheses are summarized in the 2 × 2 contingency table found in Table 9.1.

TABLE 9.1 2 × 2 Contingency Table of Type I and Type II Errors

	H_0 *True*	H_0 *False*
Reject H_0	Type I Error	Correct Rejection
Do not reject H_A	Correct Nonrejection	Type II Error

Z Test

Having discussed the abstract theoretical aspects of hypothesis testing, we are now ready to turn to an actual application. To this end we reuse the bovine data (summarized below) from the discussion that has preceded, and offer the accompanying statistical formula, called a *Z test*. As denoted by the formula, the only way we can use a Z test is when the population mean, the population standard deviation, the sample mean, and the sample size values are all known.

What the Z test enables us to determine is if the difference between the two means (μ: mu and $\bar{x}$: *x* bar) is statistically significant. As the formula states, we first subtract the population mean from the sample mean. The population standard deviation is then divided by the square root of the sample size, once again referred to as the standard error of the mean. Recall from the previous chapter that, simply stated, the standard error of the mean is a standardized measure of the amount of variability and potential sampling error found around the mean of a given data set. In terms of how it is calculated, the format of the standard error of the mean differs and is dependent upon the type of test being performed. Nevertheless, all of the statistical tests in this chapter use a form of the standard error of the mean in the denominator portion of the given formula. Finally, we simply divide the resultant numerator value by the resultant denominator value to get what is called a *Z obtained.*

The Z-obtained value is compared against a second value called the *Z critical*. In many ways, Z-obtained values are similar to Z scores discussed in Chapter 6. That is, a Z-obtained value represents how much of a standardized difference there is between the two means compared. We define Z-critical values in a moment. However, let's first do an actual Z-test calculation—as completed below.

Since only two-tailed tests are presently considered, the Z-obtained value is put into brackets, which indicates that we are treating it as an absolute value with no regard to sign. Thus the final answer is $|-10|$ or 10. Near the end of this chapter (in a separate, optional section) we discuss the difference between two-tailed and one-tailed statistical tests.

Z Test	Holstein Data
$\dfrac{\bar{x} - \mu}{\sigma/\sqrt{n}} = \text{Z Obtained}$	$\mu = 10$ $\bar{x} = 7$ $\sigma = 1.5$ $n = 25$

Recall that $\mu = 10$ represents the national average for all cows in terms of the number of times they touch the electric fence; the σ represents the standard deviation for this population. The $\bar{x} = 7$ represents the Holstein average and was determined from a sample of 25. Thus,

$$\frac{7 - 10}{1.5/\sqrt{25}} = \frac{-3}{1.5/5} = \frac{-3}{.3} = \boxed{|-10|} = \text{Z Obtained}$$

Having calculated the Z obtained, we can now turn to defining what a Z-critical value is and how it is determined. Quite simply, exactly like the Z-critical values used to construct confidence intervals, we use one of two values: 1.96 (alpha = .05) and 2.58 (alpha = .01). Further, while the manner in which they are derived is exactly the same (refer back to Chapter 8 for a more complete discussion), the current application is slightly different. Figure 9.1 represents a Z-score distribution with the corresponding Z values of 1.96 and 2.58 located on it. In reference to Z tests, what this distribution informs us is that an obtained value must fall outside of the area represented by the critical values for there to be a significant difference. That is, the closer an obtained value is to the mean (0), the more likely that the difference is a result of sampling error. Conversely, the farther the obtained value is from the mean, especially scores beyond plus or minus 1.96, the more likely there is an actual difference.

Stating this in slightly different terms, the obtained value (treated as an absolute) must be larger than the critical value for us to conclude that there is a statistically significant difference. For instance, if the obtained value is greater than 1.96, then it is significant at the .05 level. Further, this tells us that we are 95% confident that there is an actual difference between the two means; alternatively, there is a 5% chance we have committed a Type I Error and rejected the null hypothesis when it is in fact true. Moreover, the exact same thing is said when the Z-obtained value is greater than 2.58, except that now we are 99% confident that there is a difference with a corresponding 1% chance that we have committed a Type I Error.

Returning to our bovine example with the Z obtained of 10 (alpha level = .01), we obviously reject the null hypothesis and accept the research hypothesis, and conclude that there is a significant difference. After all, the obtained value of 10 is much larger than 2.58. Further, an obtained value of 10 is also significant at the .05 alpha level; how-

FIGURE 9.1 Z-Score Distribution

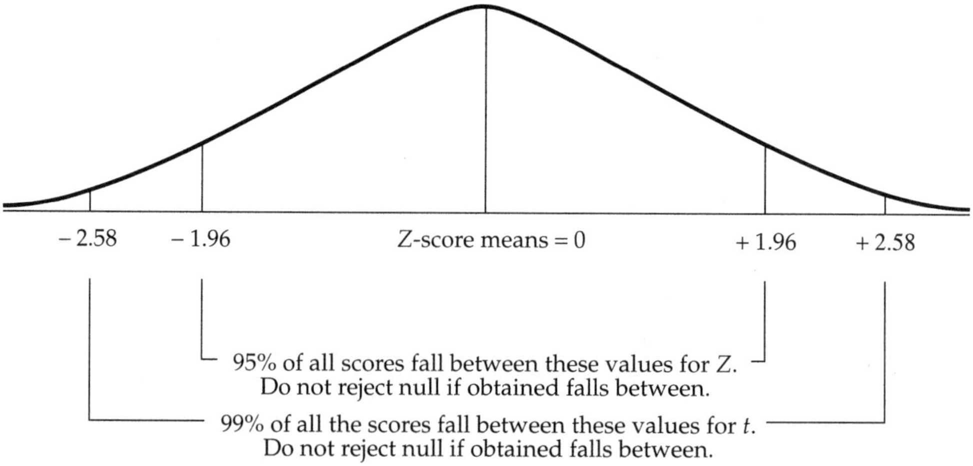

ever, since we are even more confident at the .01 level that there is a statistically significant difference, it is obviously more desirable to reject the null at the second level (.01).

In other words, Holsteins are far less likely to touch an electric fence than all other breeds of cows combined. Moreover, while there is a 1% chance that this conclusion is in error, more importantly, we are 99% confident that it is in fact true. Thus, with an electric fence as proof, we conclude that Holsteins (as a group) are apparently more intelligent than most cows.

Using a variant of the bovine cartoon to give another example of a Z test, let's say that instead of Holstein's we sample 9 Jersey cows and find that their daily average for touching the electric fence is $\bar{x} = 10.5$. This information along with the appropriate calculations are found below. Also found below is a preset alpha level of .05 with the noted corresponding Z-critical value of 1.96. Although obtained values are typically reported as either significant at the level found or not significant, to simplify our discussion we have preset the alpha level in this and subsequent tests.

<u>Jersey Data</u>

$\mu = 10$

$\bar{x} = 10.5$

$\sigma = 1.5$

$n = 9$

$$\frac{10.5 - 10}{1.5/\sqrt{9}} = \frac{.5}{1.5/3} = \frac{.5}{.5} = \boxed{1}$$

$\alpha = .05$, Z critical $= 1.96$

With the Z-obtained value of 1 in this case being smaller than the Z-critical value of 1.96, we accept the null hypothesis (reject the research hypothesis) and conclude that, statistically speaking, there is not a significant difference between the two groups. In other words, although Jerseys may appear to touch the electric fence more often than the average for all breeds of cows, this difference is not statistically significant. In sum, in terms of touching the electric fence, Jerseys are apparently no more or less intelligent than the average for all cows.

Dependent-Sample *t* Test

Sometimes when a researcher is hypothesis testing she or he may have access to a population mean but not know the accompanying standard deviation value. In cases such as this, a different type of test is used: a *dependent-sample t test*. The following is the actual formula.

$$\text{Dependent-Sample } t \text{ Test} = \frac{\bar{x} - \mu}{s/\sqrt{n}} = t \text{ Obtained}$$

Not only does this new test use a *sample standard deviation* in place of the population standard deviation found in the previous Z test, but (as you may have guessed by the test's name) it also requires the use of a *t* distribution. More specifically, the *t* distribution is used to determine critical values of *t* which are compared to obtained values of *t* to judge if there is a significant difference. The procedures to determine which *t*-critical value is used are the same ones employed to determine *t*-critical values used to construct confidence intervals. Once again, the *t*-critical value is determined by two

CARTOON 9.2

pieces of information: the alpha level and the degrees of freedom ($n - 1$).

To apply what has been discussed thus far, we offer Cartoon 9.2. The intent of this cartoon is not to offend any of the Greeks using this text, but rather just to have a little fun using statistics. After all, the first author is a former fraternity member.

Let's say our current research interest is the amount of money spent per month on clothes by different types of male students at "All-American U"—this is the population. From a previous questionnaire administered during registration to all students at All-American U we found that male students spend an average of $50.00 per month for clothes; however, this study neglected to calculate the corresponding standard deviation. We want to know if there is a difference between the amount of money spent per month for clothing by the general male population versus fraternity members (Bob and all those Polo shirts, but then again, he can spell his name backwards and get the same result). Thus, the null hypothesis states that there is no difference between the amount of money spent by all male students versus fraternity members at All-American U, whereas the research hypothesis states that there is a difference.

To test this, we draw a random sample of 25 fraternity members and find that they spend an average of $60.00 per month on clothing, with a sample standard deviation of $20.00. This information along with the actual *t* test is summarized below. Also found is a preset alpha level of .02 with its corresponding *t*-critical value of 2.492. This value, once again, is found in Appendix 2 using 24 degrees of freedom ($n - 1$; $25 - 1 = 24$ df) and the preset alpha level of .02.

Clothing Expenditures for Fraternity Members Versus the General Male Population at All-American U

$\mu = \$50.00$
$\bar{x} = \$60.00$
$s = \$20.00$
$n = 25$

$$\frac{\bar{x} - \mu}{s/\sqrt{n}} = \frac{60 - 50}{20/\sqrt{25}} = \frac{10}{20/5} = \frac{10}{4} = \boxed{2.5}$$

$\alpha = .02$, t critical $= 2.492$; t obtained $= 2.5$

With the *t* obtained of 2.5 being larger than the *t*-critical value of 2.492, we obviously conclude that there is a statistically significant difference between the two groups.

Thus, we reject the null hypothesis and accept the research hypothesis. Further we are 98% confident that we have correctly rejected the null hypothesis with a converse 2% chance that we have incorrectly rejected it when in fact it is true—a Type I Error. In total, we conclude that fraternity members at All-American U spend, on the average, more money on clothes than the general male population attending the university.

For our next example of a dependent-sample *t* test, we use the same population information previously gathered at All-American U concerning expenditures on clothing ($\mu = \$50.00$). Instead of fraternity members, however, we are now interested in the average amount of money spent on clothes by male athletes—typical attire being T-shirts and sweatshirts turned inside out to get a second day's wear—versus the overall average for all male students. We sample 16 male athletes and find that they spend an average of \$45.00 a month on clothing, with a sample standard deviation of \$10.00. With a preset alpha level of .05 and 15 degrees of freedom, the *t*-critical value for this example is 2.131 (Appendix 2). All of this information and the appropriate calculations are presented below.

Clothing Expenditures for Male Athletes Versus the General Male Population at All-American U

$\mu = \$50.00$

$\bar{x} = \$45.00$

$s = \$10.00$

$n = 16$

$$\frac{\bar{x} - \mu}{s/\sqrt{n}} = \frac{45 - 50}{10/\sqrt{16}} = \frac{-5}{10/4} = \frac{-5}{2.5} = \boxed{|-2|}$$

$\alpha = .05$, t critical $= 2.131$; t obtained $= |-2|$ or 2

With a *t*-obtained value of 2 being less than the *t*-critical value of 2.131, we accept the null hypothesis and reject the research hypothesis. Therefore, we conclude that there is not a statistically significant difference between the amount of money spent on clothing by male athletes versus the general male population at All- American U. Then again, there might be something said for turning soiled sweatshirts inside out if one is looking to save a little money.

Independent-Samples *t* Test

Similar to the construction of confidence intervals, typically no information about the population is available to the researcher when he or she sets out to test hypotheses. And as you have probably already guessed, such situations call for yet another type of *t* test. One that is often used, which we discuss now, is called the *independent-samples t test*. The actual formula is found below. Please note that part of the formula, s^2 pooled, is presented separately to simplify the actual calculations.

Although you should be familiar with all the symbols found within the formula, before proceeding it is still a good idea to clarify what they represent and briefly review them. While the standard error of the mean no longer appears in the same form as it has thus far, the denominator portion of the formula still represents it. That is, mathematically, the denominator portion of the formula is the standard deviation divided by the square root of *n*. The reason that the formula appears in this form, however, is because seldom, if ever, are two samples drawn that have the exact same sample size and/or standard deviations. As such, the denominator is mathematically a weighted standard error of the mean. In other words, similar to a weighted mean, an independent-samples t test takes into account different standard deviation values and sample sizes.

The whole formula, including the s^2 pooled portion, has only three basic symbols: (1) x for sample means, (2) s for sample standard deviations, and (3) n for sample sizes. Since this formula tests to see if there is a difference between two independent sample means, also found for each symbol is a subscript for which group the information belongs to. In other words, the subscript 1 refers to sample one—group 1; conversely, the subscript 2 refers to sample two—group 2.

Independent-Samples t Test	s^2 Pooled Portion of Formula
$\dfrac{\bar{x}_1 - \bar{x}_2}{\sqrt{s^2 \text{ pooled } (1/n_1 + 1/n_2)}}$	$\dfrac{(n_1 - 1)\, s_1^2 + (n_2 - 1)\, s_2^2}{n_1 + n_2 - 2}$

We are now ready to turn to the formula's actual calculation. To this end, we refer back to Cartoon 9.2 as a backdrop for our first example. Let's say our current research interests are the number of organized social events per month that fraternity members attend versus non-fraternity men at All-American U. We get two lists from the registrar's office, one for fraternity members and one for non-fraternity (all other) men at the campus. From this, we draw a sample from each of the groups, two independent samples, and find the following. Also listed is a preset alpha level of .01. The actual calculations using the above formula are found below the data sets.

Fraternity Members Group 1	Non-Fraternity Men Group 2	
$\bar{x}_1 = 15$	$\bar{x}_2 = 11$	Average Number of Social Events Attended per Month
$s_1 = 3$*	$s_2 = 4$	Standard Deviation of Events
$n_1 = 33$	$n_2 = 29$	Sample Size of Each Group

(alpha = .01)

* Why do you think the standard deviation for fraternities ($s = 3$) is smaller than that for non-fraternity members ($s = 4$)? Since fraternity members are a more homogeneous (similar) group than simply everyone else (a more heterogeneous group), one expects a smaller amount of variation in this group.

$$\frac{15 - 11}{\sqrt{\dfrac{(33 - 1)\,(3)^2 + (29 - 1)\,(4)^2}{33 + 29 - 2}\left(\dfrac{1}{33} + \dfrac{1}{29}\right)}}$$

$$= \frac{4}{\sqrt{\dfrac{(32)\,(9) + (28)\,(16)}{60}\,(.0303 + .0344)}}$$

$$= \frac{4}{\sqrt{\frac{288 + 448}{60}(.0647)}} = \frac{4}{\sqrt{(12.266\bar{6})(.0647)}} = \frac{4}{.8908} = \boxed{4.4903}$$

While the above calculations give us a *t* obtained of 4.4903, this answer alone does not tell us if the two sample means are significantly different. To make this determination, we obviously must decide what the *t*-critical value is. For this type of *t* test, however, we use a slightly different operation than the previous, $n - 1$. Since we are dealing with *two* independent samples, this must be taken into account when determining the *t*-critical value. In other words, since two samples are used, correspondingly, two degrees of freedom must be controlled.

To do this, we take the sum of both samples and subtract the value of 2 from it; $n + n - 2 =$ *t*-critical df or (for this example) $29 + 33 - 2 = 60$. (Please note that in the denominator of the *s*-pooled portion of the formula one finds exactly the same calculations. Some of you may prefer to simply use this portion of the formula to calculate the degrees of freedom for this type of *t* test.) With 60 degrees of freedom and the alpha level preset at .01, we look in Appendix 2 and find that the *t*-critical value for this example is 2.66.

Now we can finally determine if there is a significant difference. With a 4.4903 *t* obtained being larger than the 2.66 *t* critical, we conclude that there is a significant difference between the two groups; thus, we reject the null hypothesis and accept the research hypothesis. With the alpha level at .01, we also conclude that there is a 1% chance that we have committed a Type I Error; alternatively, we are 99% confident that we have correctly rejected the null hypothesis. In other words, fraternity members apparently attend significantly more organized social events per month than non-fraternity members. Given the number of private mixers and other types of parties that fraternities sponsor, this finding is really not that surprising.

For our next example of an independent-samples *t* test, let's say we now want to know if there are differences in the dating behaviors of fraternity versus non-fraternity men. That is, during a school year does one group go out on more dates with different women than the other? Once again, we draw two independent samples and find the following with accompanying calculations.

Fraternity Members Group 1	Non-Fraternity Men Group 2	
$\bar{x}_1 = 8$	$\bar{x}_2 = 7$	Average Number of Dates per School Year
$s_1 = 1.5$	$s_2 = 3$	Standard Deviation of Dates
$n_1 = 19$	$n_2 = 21$	Sample Size of Each Group

(alpha = .05)

$$\frac{8-7}{\sqrt{\frac{(19-1)(1.5)^2+(21-1)(3)^2}{19+21-2}\left(\frac{1}{19}+\frac{1}{21}\right)}}$$

$$=\frac{1}{\sqrt{\frac{(18)(2.25)+(20)(9)}{38}(.0526+.0476)}}$$

$$=\frac{1}{\sqrt{\frac{40.5+180}{38}(.1002)}}=\frac{1}{\sqrt{(5.8026)(.1002)}}=\frac{1}{.7625}=\boxed{1.3114}$$

With a *t* obtained of 1.3114, we need to compare it to a *t*-critical value that has 38 degrees of freedom with an alpha level of .05. Referring once again to Appendix 2 (as some of you probably already have), we find that although there are *t*-critical values for 30 df and 40 df, a *t* critical for 38 df is not listed. To simplify matters in cases such as this, we ask that you always take the larger, more conservative value associated with the lesser degrees of freedom. For this example, then, we use 30 df. With the alpha set at .05, the corresponding *t*-critical value is 2.042. (To give a few more examples, let's say we have 100 df; in this case we would use the lesser value of 60 df. Or let's say we have 55 df; in this case we would use 40 df.)

Since the *t* obtained (1.3114) is smaller than the *t*-critical value (2.042) we conclude that, statistically speaking, there is not a significant difference between the two groups. Thus, we accept the null hypothesis and reject the research hypothesis. In other words, there is no difference in the dating behaviors of fraternity men versus non-fraternity members men.

Difference (Matched or Paired) *t* Test

The *difference t test*, also referred to as a matched or paired *t* test, is somewhat different than the *Z* test and *t* tests explored thus far. What distinguishes a difference *t* test is that whereas the previous tests used two separate group means, this test uses two different means obtained from the same group. That is, a difference *t* test typically uses a single group of people, derives two means from them, and then tests to see if there is a difference between them. Obviously, however, the two means compared are measurements of the same variable. The manner in which this is typically applied (there are other applications, e.g., twins research) is to give a pretest, then some sort of treatment, followed by a post-test to determine if the treatment brought about a measurable change.

As usual, the easiest way to clarify what has been abstractly stated thus far is with another cartoon problem. Cartoon 9.3 serves as a backdrop to the last few problems discussed in this chapter. To begin with and reflecting on the cartoon, let's say we want to know if listening to heavy metal music increases or decreases the number of pages a

CARTOON 9.3

given student reads nightly. We randomly select seven ($n = 7$) individuals attending All-American U and measure the number of pages they read per school night listening to no music (the pretest). Next, we have these same individuals listen to heavy metal music (the treatment) while they are reading. As they listen to the music, we also measure the number of pages they read per school night (the post-test). Our findings are reported in Table 9.2. Found below is the formula (and its sub-parts) used for calculating a difference t test.

TABLE 9.2 Heavy Metal/Reading Data

	Number of Pages Read	
Individual	*No Music*	*Listening to Heavy Metal*
1	34	37
2	31	43
3	18	27
4	36	33
5	31	39
6	30	30
7	38	41

Difference *t*-Test Formula

$$\frac{\bar{d}}{s_d/\sqrt{n}}$$

where

$$\bar{d} = \frac{\Sigma d}{n} \text{ and } s_d = \sqrt{\frac{\Sigma d^2 - (\Sigma d)^2/n}{n - 1}}$$

Since there are new symbols found in this formula, before doing any calculations we must first explain what it actually asks us to do. To begin with, and as you may have guessed, since this is a difference *t* test, it asks us to determine the difference between two sets of scores. In mathematical terms, a "difference" ultimately refers to subtraction.

Referring to the actual formula, in the numerator portion is a *d* with a bar above it ($\bar{d}$). Symbolically similar to a sample mean, this formula sub-part asks us to calculate the mean difference of the scores. To accomplish this, we simply take each individual's first score and subtract their corresponding second score from it (e.g., individual 1: $34 - 37 = -3$). Then, as the formula notes, all the resultant scores are added together and then divided by the sample size.

Turning to the symbol of *s* with *d* subscript (s_d), this represents the standard deviation of difference. Beginning with the left symbol in the numerator, this informs us to take the sum of the differences squared. That is, each individual difference score is squared and then summed. The right-hand portion of the denominator, alternatively, tells us to take the sum of the differences (the same value used in the mean calculation), square the resultant value, and then divide it by the sample size ($n = 7$). In the denominator is the same operation that is associated with a regular sample standard deviation: $n - 1$, or the sample size minus one.

With these instructions in mind, let's go ahead and apply them to the above data set. Although all the calculations are completed below, since this is our first time through this formula we walk you through each of the operations. First, in addition to the original data set we have added a column for the individual differences and a column for each of these values squared. Below each new column is found its sum; each

TABLE 9.3 Intermediate Steps for Heavy Metal Difference Test

Individual	*No Music*	*Listening to Heavy Metal*	*d (Difference)*	*d^2 (Difference Squared)*
1	34	37 (34–37)	−3	9
2	31	43 (31–43)	−12	144
3	18	27 (18–27)	−9	81
4	36	33 (36–33)	+3	9
5	31	39 (31–39)	−8	64
6	30	30 (30–30)	0	0
7	38	41 (38–41)	−3	9
			−32	316

of these values is required in subsequent calculations. Please note that when the differences are summed, we must be especially careful when adding negative values or the resultant answer will be wrong (e.g., $-3 + -12 = -15$, or $-6 + 3 = -3$). This is why for each difference score we have clearly designated it as a negative or positive value.

Next we calculate the mean difference. Once again, this is obtained by taking the sum of the differences and dividing it by the sample size: $-32/7 = -4.5714$. After this, we calculate the standard deviation of differences: 5.3183. Having derived these two initial values, combined with the sample size value, we simply place all of these values into their appropriate locations in the actual formula and get a t obtained of -2.1665. (In other words, both the mean difference and the standard deviation of differences must first be separately calculated before the t obtained can be derived.)

To determine the t critical, as with previous t tests, we preset an alpha level (.05 for this example) and determine the degrees of freedom. Since there are seven individuals in this example, we simply take $n - 1 = \text{df}$ or $7 - 1 = 6$ df. [Note that even though we have two measurements of seven, 14 total, since just one variable is measured (pages read), our n is the actual number of subjects involved in the experiment.] This gives us a t-critical value of 2.447. Once again, all of the above discussion is summarized in Table 9.3 (the intermediate calculations) and below with the final calculations.

Mean Difference

$$\bar{d} = \frac{-32}{7} = \boxed{-4.5714}$$

Standard Deviation of Differences

$$s_d = \sqrt{\frac{d^2 - (d)^2/n}{n-1}} = \sqrt{\frac{316 - (-32)^2/7}{7-1}} = \sqrt{\frac{316 - 1024/7}{6}}$$

$$= \sqrt{\frac{316 - 146.2857}{6}} = \sqrt{\frac{169.7143}{6}} = \sqrt{28.2847} = \boxed{5.3183}$$

Difference t test

$$\frac{\bar{d}}{s_d/\sqrt{n}} = \frac{-4.5714}{5.3183/\sqrt{7}} = \frac{-4.5714}{5.3183/2.6457} = \frac{-4.5714}{2.0101} = \boxed{|-2.2742|}$$

alpha = .05, df = 6 (7 − 1 = 6); thus t critical = 2.447

The above calculations complete, we are now ready to determine if there is a statistically significant difference. Since the t obtained value of $|-2.2742|$ is smaller than the t-critical value of 2.447, we conclude that there is not a significant difference; we accept the null and reject the alternative hypothesis. Thus, we conclude that listening to heavy metal music versus no music has no appreciable effect on the number of pages read nightly by these seven research subjects.

If we were actually researching this topic, we would probably still report that listening to heavy metal music does appear to increase the number of pages read; however, the increase is not statistically significant. One way that we could probably demonstrate this relationship to be statistically significant is to increase our sample size. Whenever sample sizes are increased, mathematically it makes the denominator value smaller (the square root of n), and thus the t obtained often becomes larger. Further, when the sample size increases the corresponding t-critical value becomes smaller, which makes it easier to find significance. Both of these factors make it easier to reject the null hypothesis. Of course, this also assumes that if more subjects were tested, the relationship would continue in this manner. A second, easier way that this relationship can be demonstrated as significant is discussed in Box 12 using a one-tailed t-critical value.

To hopefully ensure that everyone clearly understands how to calculate a difference t test, we offer one last example based upon Cartoon 9.3. Let's say we replicate the previous experiment with 10 individuals; however, instead of having them listen to

TABLE 9.4 Intermediate Steps for Barry Manilow Difference Test

Individual	*No Music*	*Listening to Barry Manilow*	*d (Difference)*	*d^2 (Difference Squared)*
1	33	23	+10	100
2	29	31	−2	4
3	19	12	+7	49
4	38	29	+9	81
5	33	20	+13	169
6	32	30	+2	4
7	39	42	−3	9
8	25	18	+7	49
9	22	21	+1	1
10	37	24	+13	169
			57	635

heavy metal music, we have them listen to Barry Manilow recordings. Again we measure the number of pages read per night not listening to any music (pretest), and then the number of pages read listening to Barry Manilow music. Table 9.4 represents our findings and the intermediate calculations. Found below are the appropriate formulas and their corresponding calculations. Again we have preset the alpha level at .05.

Mean Difference

$$\bar{d} = 57/10 = \boxed{5.7}$$

Standard Deviation of Differences

$$s_d = \sqrt{\frac{d^2 - (d)^2/n}{n - 1}} = \sqrt{\frac{635 - (57)^2/10}{10 - 1}} = \sqrt{\frac{635 - 3249/10}{9}}$$

$$= \sqrt{\frac{635 - 324.9}{9}} = \sqrt{\frac{310.1}{9}} = \sqrt{34.4555} = \boxed{5.8698}$$

Difference *t* test

$$\frac{\bar{d}}{s_d/\sqrt{n}} = \frac{5.7}{5.8698/\sqrt{10}} = \frac{5.7}{5.8698/3.1622} = \frac{5.7}{1.8562} = \boxed{3.0707}$$

alpha = .05, df = 10 (10 − 1 = 9); thus *t* critical = 2.262

With a *t* obtained of 3.0707 and the *t*-critical value of 2.262, we conclude that listening to Barry Manilow music has a statistically significant affect on the number of pages. Specifically, and reflecting the raw numbers in the data set, listening to Barry Manilow music significantly reduces the number of pages read per night. This is in direct contrast to the previous example, where listening to heavy metal music led to more pages being read. However, recall that this relationship was not found to be statistically significant. Regardless, since this is significant, we reject the null hypothesis and accept the research hypothesis. As such, we are 95% confident that we have correctly rejected the null hypothesis; however, there is still a 5% chance of a Type I Error. In sum, don't listen to Barry Manilow music if you have a lot of pages to read in a short period of time.

Box 12

In an attempt to simplify the material presented in this chapter, we preset all alpha levels and only use two-tailed critical values to determine significance. When social scientists are undertaking actual research projects, however, they seldom preset levels of significance and often use *one-tailed critical values.* In both cases, there are researcher advantages for not using the simplified format that we have followed.

While we have preset the alpha levels, actual researchers almost always report the level to which the *t* (or *Z*) obtained is significant. In the Barry

Manilow example, for instance, we obtained a *t* value of 2.9634 and noted that since it is larger than the *t*-critical value of 2.262 (alpha = .05, 9 df), it is significant at this level. While this test is significant at the .05 level, moreover, it is also significant at the .02 level, but it is not significant at the .01 level.

To demonstrate, please refer to Appendix 2 and follow along the row heading of 9 degrees of freedom: the df for this problem. Note that the *t* obtained value of 2.9634 is larger than the *t*-critical values at the both .05 and .02 alpha levels but it is not larger than the *t*-critical value at the .01 alpha level: 3.250. Thus, this problem is more appropriately reported as statistically significant at the .02 alpha level. With this example, obviously it is to a researcher's advantage to report this finding significant at the .02 level rather than the .05 level. After all, we are more confident at this level because there is a diminished possibility of a Type I Error.

Further, in many research reports and journals this finding would be reported with the following symbol: $p < .02$. Not only does this inform us that significance is obtained at the .02 level, but—further—the probability that a Type I Error has been committed is less than 2%. Thus *p* (probability) < (less than) .02. What if in another example we were presented with $p < .01$? This inform us that the probability that a Type I Error has been committed is less than 1%. Sometimes you may see the symbolism of $p > .05$. As you may have guessed, this simply states that the probability of a Type I Error is greater than .05, and thus it is not significant at this level. For almost all social scientists, the alpha level of .05 is the minimum to report a given relationship as statistically significant.

As noted before, often a researcher's whole project is contingent upon her or his demonstrating statistical significance. If researchers can hypothesize the direction of the relationship prior to testing it, they can use a one-tailed *t*-critical value. This is symbolically represented in the research hypothesis as $\bar{x} > \bar{x}$ or $\bar{x} < \bar{x}$ instead of the previously utilized $\bar{x} \neq \bar{x}$. When the use of a one-tailed value is possible, the corresponding *t*-critical values are always smaller than their two-tailed counterparts; thus, it is often much easier to find statistical significance using one-tailed values.

In the previous example (number of pages read by subjects listening to no music versus heavy metal music), we were unable to find statistical significance. If, however, we had some previous reason to believe that listening to heavy metal music might increase the number of pages read (such as previous research that had demonstrated this or a similar relationship), we could have treated this as a one-tailed *t* test. That is, instead of using the critical *t* value of 2.447 (it resulted in our concluding that there was not a significant difference), we could use the *t*-critical value of 1.943 (look to the one-tailed column heading with 6 df, alpha level = .05). In other words, had we hypothesized the direction of this relationship, we would have used the *t*-critical value of 1.943. And with a *t*-obtained value of $|-2.1665|$, we would have concluded this relationship to be statistically significant at the .05 alpha level. Then again, those of you who listen to this type of music probably already knew this.

Chapter Summary and Conclusions

This chapter has explored techniques of hypothesis testing. To this end, specifically the techniques of the Z test, dependent-sample t test, independent-samples t test, and difference t tests were presented. All four of these tests made comparisons of some form of group means to determine if there were statistical difference between them. The selection of which of these tests is used is dependent upon the information available to the researcher and/or the manner in which she or he collected the data. In variant forms, the underlying assumptions of these tests and the actual tests themselves are reapplied in the statistical tests found in the remaining chapters.

Key Terms To Remember

Bivariate Relationships	Z Tests
Multivariate Relationships	Dependent-Sample *t* Test
Null Hypothesis	Independent-Samples *t* Test
Research Hypothesis	Paired *t* Test
Type I Errors	One-Tailed *t*-Critical Values
Type II Errors	Two-Tailed *t*-Critical Values

Practice Exercises

1. Referring, once again, to the bovine data and the electric fence, let's say that we want to know how intelligent the breed of Brown Swiss are—the cows used to make Miss Swiss Hot Chocolate. We sample 9 Brown Swiss cows and find they touch the electric fence an average of 13 times per day. The combined information gives us the following:

$$\mu = 10; \quad \bar{x} = 13; \quad \sigma = 1.5; \quad n = 9$$

   a. Using the appropriate test (alpha = .05), is the Brown Swiss mean statistically different than that of the national average?
   b. What does this tell us in the context of the null and research hypotheses?

2. Let's say that we now want to now if the breed of Guernsey cows are more or less intelligent than the national average. We sample 25 Guernsey cows and find that they touch the electric fence an average of 10.5 times per day. The combined information gives us the following:

$$\mu = 10; \quad \bar{x} = 10.5; \quad \sigma = 1.5; \quad n = 25$$

   a. Using the appropriate test (alpha = .05), is the Guernsey mean statistically different than that of the national average?
   b. What does this tell us in the context of the null and research hypotheses?

3. Referring back to Cartoon 9.2, let's say that instead of fraternity members we are now interested in sorority members and the amount they spend monthly on clothes. Once again, from a questionnaire administered at All-American U we know the average amount of money spent by all women is \$60.00. We randomly sample 36 sorority members and find that they spend an average of \$75.00 per month on clothes, with a corresponding standard deviation of \$24.00. The combined information gives us the following:

$$\mu = \$60.00; \quad \bar{x} = \$75.00; \quad s = \$24.00; \quad n = 36$$

   a. Using the appropriate test (alpha = .05), are the averages statistically different?
   b. What does this tell us in the context of the null and research hypotheses?
   c. In light of the previous fraternity example and assuming that Greek members come from more privileged backgrounds, does gender or SES best explain the spending behaviors of Greek versus non-Greek members? Why?

4. Let's say that we are interested in the number of alcoholic beverages consumed per week by students at different universities. We draw *two independent samples* of students: (1) collected at All-American U and (2) All-American State University. We find the following:

$$\bar{x}_1 = 18 \qquad \bar{x}_2 = 14$$
$$s_1 = 2.5 \qquad s_2 = 1.5$$
$$n_1 = 15 \qquad n_2 = 12$$

   a. Using the appropriate test (alpha = .001), is there a statistically significant difference between the two groups in terms of weekly consumption of alcoholic beverages?

5. Similarly, let's say that we are interested in the number of parties attended per semester by students at the above universities: once again, 1) collected at All-American U. and 2) All-American State University. We find the following:

$$\bar{x}_1 = 25 \qquad \bar{x}_2 = 21$$
$$s_1 = 4 \qquad s_2 = 3$$
$$n_1 = 15 \qquad n_2 = 12$$

   a. Using the appropriate test (alpha = .02), is there a statistically significant difference between the two groups in terms of the number of parties they attend per semester?

6. Let's now say we want to know if listening to country music increases or decreases the number of pages read per night. We sample 9 research subjects and find the data listed in Table 9.5.
   a. Using the appropriate test (alpha = .05), does listening to country music have a statistically significant effect on the number of pages read per night? What sort of effect does it have?

7. Finally, let's say we want to know if listening to the blues increases or decreases the number of pages read per night. We sample 12 research subjects and find the data listed in Table 9.6.

   Using the appropriate test (alpha = .05), does listening to the blues have a statistically significant effect on the number of pages read per night? What sort of effect does it have?

TABLE 9.5 Country Music/Reading Data

	Number of Pages Read	
Individual	*No Music*	*Listening to Country Music*
1	33	21
2	31	32
3	18	16
4	36	33
5	31	24
6	30	22
7	38	41
8	24	15
9	23	12

TABLE 9.6 Blues/Reading Data

	Number of Pages Read	
Individual	*No Music*	*Listening to The Blues*
1	23	31
2	21	42
3	28	26
4	31	43
5	18	34
6	20	32
7	32	51
8	34	25
9	33	22
10	29	31
11	23	22
12	33	36

Chapter 10

Simple Regression and Correlation

This text has largely explored univariate (single-variable) statistical techniques. In the last chapter we began to consider bivariate (two-variable) statistical techniques. This was in the context of comparing two groups' means to determine if a statistically significant difference existed (e.g., breed of cow and number of times it touches the electric fence). Although the techniques discussed to this point are all important and meaningful ways of analyzing variables, they tell us very little about the form or potential strength of a relationship between two variables.

Further, in the last few chapters we have been primarily concerned with relationships between nominal and interval-ratio or ordinal variables. The statistical techniques in this chapter deal with bivariate relationships between two interval-ratio variables in terms of whether, and how strongly, they are associated together. Specifically, this chapter first explores a simple graphing technique called a scattergram that is used to illustrate relationships between two variables. Next we delve into the important statistical techniques of simple linear regression, Pearson's r, and a newly introduced t test of significance. Similar techniques for ordinal and nominal variables are also considered. Before proceeding, however, a brief review of independent and dependent variables and a discussion of causality is warranted.

Independent/Dependent Variables and Causality

Almost all research is based on the notion of causality. In the last chapter we implicitly introduced the idea of causality in terms of whether a given variable was seen as bringing about a measurable change in another group's mean. Quite simply, causality is the notion that one event or series of events causes another event or series of events to occur. In more scientific terms, one variable is seen as bringing about—causing—a change in another variable.

The specific terms used to describe this relationship between two variables are *independent (X)* and *dependent (Y) variables*. As we have previously discussed, the independent variable is seen as bringing about a change in a dependent variable. Stated slightly differently, the dependent variable is seen as a function of the independent variable. Conversely, the independent variable is seen as a determinant of the dependent variable.

A simple example of this relationship is a person's income (the dependent variable) seen as a function of years of education (the independent variable). Alternatively, the grade you receive in this class (the dependent variable) is seen as a function of the number of hours spent studying (the independent variable). In other words, years of education and hours spent studying are seen as determinants of, respectively, one's income and course grade.

While both of these examples make sense intuitively , unfortunately, due to extraneous (unaccounted for) factors that are often not measured, we are never able absolutely to conclude causality. That is, while the relationship of hours spent studying as a determinant of the grade you receive in this class makes sense, potential other factors (such as previous math classes taken, type of school one attends, whether one is employed, has children or other social obligations while attending school) can all potentially effect one's course grade. In other words, while two variables may be associated together—they tend to occur together—it does not follow automatically that one causes the other to change.

Moreover, as just demonstrated, most if not all variable relationships in the social sciences involve several, often numerous, independent variables. Often these "other" variables are not accounted for in a given research design. For instance, the list of potential determinants of one's grade listed above is far from complete (e.g., type of textbook and/or instructor you have while taking this course). As such, it is impossible to conclude absolutely that one variable is the cause of another. Quite simply, in the social sciences, at best, we make educated and logical inferences about independent and dependent variable relationships.

A somewhat humorous example that summarizes the above discussion is the often noted relationship between fire trucks and fire damage. In numerical terms, as more fire trucks and firefighters respond to a fire there is a strong tendency for more damage to occur. One could conclude that fire trucks (or firefighters) are the independent variable and fire damage is the dependent variable; that is, an increase in the number of fire trucks leads to increased damage. Of course, this is ridiculous. Nevertheless, one could conclude wrongly that the fire trucks are the cause of the damage. In reality, however, the larger the fire, the more fire trucks that are required to put it out; and the larger the fire the more damage it causes.

Scattergrams

One easy way to represent a relationship of association between two variables is graphically with a *scattergram*. Similar to his-

TABLE 10.1 Soda and Urination Data

Individual's Name	*Number of 12-ounce Sodas Consumed* (X)	*Number of Times Research Subject Urinated in the Following 3-Hour Period* (Y)
Rick	1	2
Janice	2	1
Paul	3	3
Susan	3	4
Cindy	4	6
John	5	5
Donald	6	5

tograms and frequency polygons, scattergrams entail the graphic charting of two variables. This is accomplished by the use of intersecting axes; the horizontal axis represents the independent variable while the vertical axis represents the dependent variable. A dot is then used to represent each measurement of the independent variable in comparison to the dependent variable.

An easier way to discuss scattergrams is with an actual example. Taking a short break from cartoon examples but still using one that is absurd but somewhat humorous is the data set found in Table 10.1 for seven research subjects. The independent variable, under the heading *X*, is the number of 12-ounce sodas consumed by these seven research subjects. The dependent variable, under the heading *Y*, is the corresponding number of times each subject urinated in the three-hour period immediately following the consumption of the sodas. To keep this presentation simple, we have rank ordered the observations of the independent variable.

Two preliminary observations are warranted. First (as some of you have probably already noticed from the data set), as the amount of soda consumed increases, correspondingly, and not surprisingly, there is an increase in the number of times the research subject visits the bathroom. Second, while this example may appear completely absurd, as an undergraduate in college the first author worked in a federally sponsored nutrition research center where similar experiments were undertaken.

To plot a scattergram with this data set we use the horizontal axis to represent the *X* variable and the vertical axis to represent the *Y* variable. Then we place a dot where the observations of the *X* and *Y* variables for each individual meet. As found in Figure 10.1, the first dot for individual 1 is found at the intersection of 1 for *X* and 2 for *Y*. The rest of the dots representing each individual are also plotted; the resultant picture is called a scattergram. Due to such a small sample size, the scattergram represented in Figure 10.1 is fairly neat in presentation. Many scattergrams, however, plot hundreds (sometimes thousands) of cases, which often results in a mirage of dots.

A picture emerges from our scattergram that allows us to infer that there is a positive linear relationship between the independent and dependent variables. Linear, quite simply, means that the relationship between these two variables resembles a straight line. That is, one of the best ways to represent the relationship between these

FIGURE 10.1 Scattergram

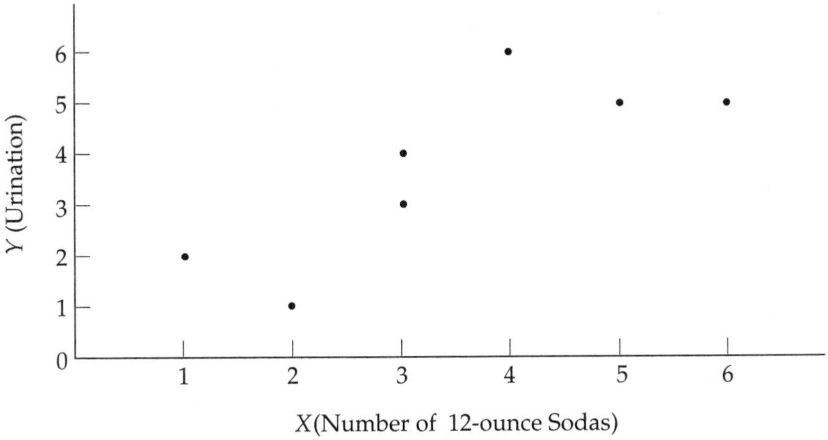

two variables is to fit a straight line through the pattern of the scattergram. In a sense, this fitted line is an average of where the actual dots occur. In the next section we do this mathematically. For now, however, we simply estimate where this line goes and get the illustration of a regression line placed over the previously constructed scattergram (Figure 10.2).

From this new depiction, one quickly notes that as the line goes from left to right, it correspondingly rises in reference to the Y axis; that is, with each increase of the X variable comes a corresponding increase in the observed Y variable. This type of relation-

FIGURE 10.2 Scattergram with Estimated Regression Line

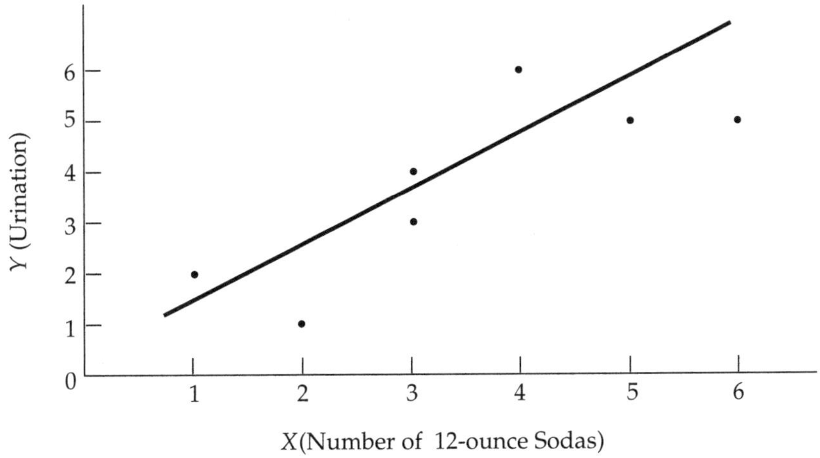

FIGURE 10.3 Curvilinear Relationship

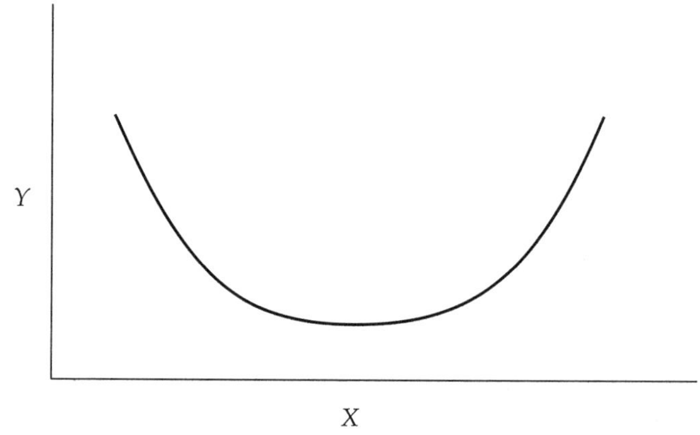

ship is referred to as a positive linear relationship. Conversely, in cases where each increase of X results in a corresponding decrease in the observed Y variable, the relationship is referred to as an inverse (negative) linear relationship. Although beyond the scope of this text, other sorts of linear relationships exist that do not follow a straight line. An example of one of these, graphically presented in Figure 10.3, is called a curvilinear relationship.

Finally, we are sometimes presented with situations where no discernible patterns emerge. That is, the dots in the scattergram are so dispersed that no lines can be fitted in any meaningful manner. In these cases, the variables are probably not associated

FIGURE 10.4 Scattergram of an Unrelated X and Y Variable Relationship

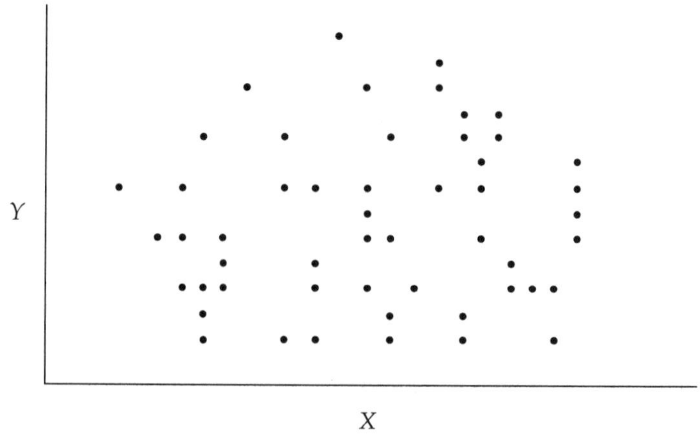

together. The scattergram found in Figure 10.4 represents an unrelated X and Y variable relationship.

Simple Linear Regression

While a scattergram gives a graphic representation—a picture—of a relationship between two variables, it does not tell us exactly where the fitted line actually belongs. We simply guessed in the above example. A precise statistical technique that does this mathematically is called a least-squares regression line, also referred to as simple linear regression. In a limited sense, this line is a mathematical average between the observed X and Y values. The actual equation to do this, found below, should look quite familiar to many of you; it is the mathematical equation for plotting a straight line. Found below the equation are the definitions for each of its parts.

Equation for a Straight Line
(Least-Squares Regression)

$$Y = a + bX$$

where

Y = *Dependent Variable*: seen as a function of (or predicted by) the independent variable (X);

X = *Independent Variable*: the dimension or characteristic that is seen as the determinant or cause of the dependent variable (Y);

b = *Slope of the Line*: rise (or drop) divided by run;

a = *Y-intercept*: where the value of $X = 0$ and the line intercepts the Y axis.

Before introducing an example that requires the actual mathematical operations to determine the slope (b) and Y-intercept (a), we will use an exact linear relationship from the physical sciences. Recall that there are examples of exact linear relationships in the physical sciences. This is not only a simple way to illustrate what the slope and intercept are but it is also gives you some practice plotting the least-squares regression line.

One simple example of an exact linear relationship is converting Celsius to Fahrenheit: two different measures of temperature. In converting these different measures of temperature it is commonly known that the slope (b) is 9/5 (1.8) and the Y-intercept (a) is 32. Stating this in slightly different terms, for each degree of Celsius added, a corresponding 1.8 increase is found on the Fahrenheit scale. Further, the value of zero (0) on the X axis (representing Celsius) is equal to 32 on the Y axis (representing Fahrenheit). Not only is this where the Y-intercept occurs, moreover, this also represents the temperature where water freezes on both scales. In increments of 10 for X we have gone ahead and calculated each score and then placed its corresponding dot on Figure 10.5. All that is left to do is to connect the dots, as is also done, which gives us a straight line that represents an exact linear relationship.

As already noted, due to unknown extraneous factors, there are probably no exact linear relationships in the social sciences. As far as that goes, there are very few in the physical sciences. As a result, we are forced to fit (calculate) a line. To assist in this dis-

FIGURE 10.5 An Exact Linear Relationship

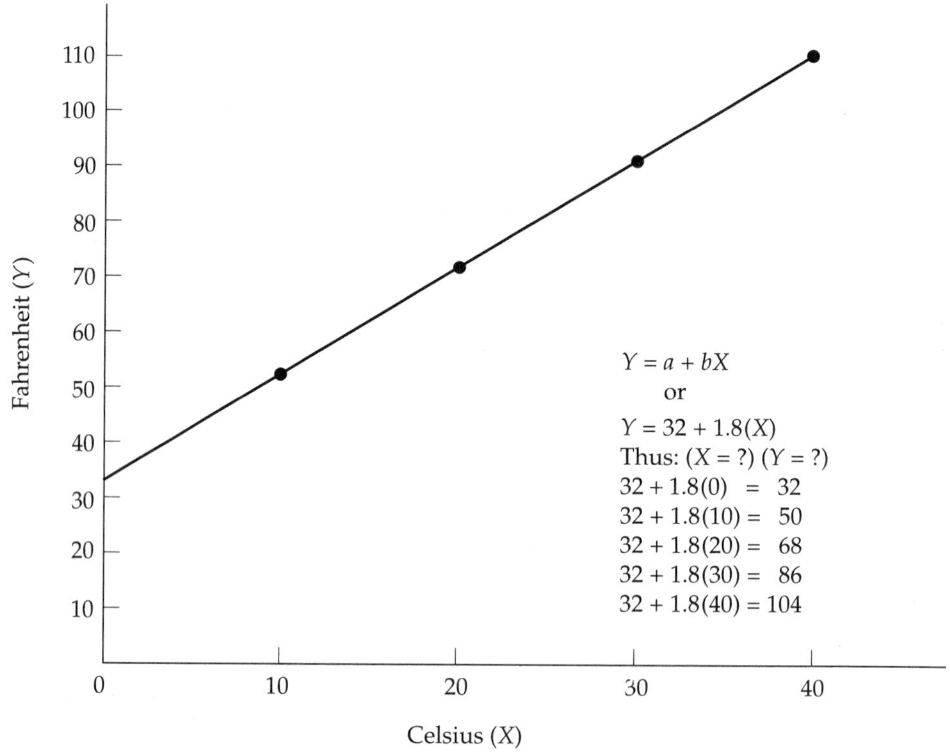

cussion we have reutilized the data set on soda and urination (Table 10.2). Also, since the relation between soda and urination is not an exact linear relationship, the formulas used to calculate the slope (b) and Y-intercept (a) are now introduced. Both of these formulas are found below.

Formulas for Calculating a and b

$$b = \frac{\Sigma XY - \frac{(\Sigma X)(\Sigma Y)}{n}}{\Sigma X^2 - \frac{(\Sigma X)^2}{n}} \qquad a = \bar{y} - b\bar{x}$$

Although there are no new symbols introduced (and all of them should be quite familiar to you), for these first calculations we walk you through the mathematical operations required for each formula. Beginning with the slope (b), the numerator portion of this formula requires us to first individually multiply each X value by each Y value and then sum these values ($\Sigma XY = 104$). Next, we take the sum of the X observations ($\Sigma X = 24$) and multiply it by the sum of the Y observations ($\Sigma Y = 26$) and find

TABLE 10.2 Replicated Soda and Urination Data

Individual's Name	*Number of 12-ounce Sodas Consumed* (X)	*Number of Times Research Subject Urinated in the Following 3-Hour Period* (Y)
Rick	1	2
Janice	2	1
Paul	3	3
Susan	3	4
Cindy	4	6
John	5	5
Donald	6	5

that (24)(26) = 624. This value, 624, is then divided by $n = 7$ (the sample size), which gives us the value of 624/7 = 89.1428. Then, as the formula requires, the value of 89.1428 is subtracted from the value of 104 (104 − 89.1428 = 14.8572). The resultant answer of 14.8572 is the solution to the numerator portion of the slope formula.

The denominator requires that each *X* value is first squared, and then these values are summed; the sum of the *X* values squared is 100. Next, the sum of all the *X* values ($\Sigma X = 24$) is squared; 24 squared = 576. This value, 576, is divided by n ($n = 7$); 576/7 82.2857, and then subtracted from the value of 100. Thus, 100 − 82.2857 = 17.7143, the denominator solution to the formula. Finally, the numerator value is divided by the denominator value, with the resultant answer being the slope; 14.8572/17.7143 = .8387 = b.

A quick observation is warranted before turning to the intercept formula. Since the answer above for slope is a positive value (.8387), this immediately informs us that there is a positive relationship between the *X* and *Y* variables. If, conversely, a negative

TABLE 10.3 Intermediate Calculations for Soda/Urination Data

	X	Y	X^2	XY
1	1	2	1	2
2	2	1	4	2
3	3	3	9	9
4	3	4	9	12
5	4	6	16	24
6	5	5	25	25
7	6	5	36	30
	24	26	100	104

value would have resulted from our calculations, the relationship between the X and Y variables would have been negative and inverse.

Turning to the calculation of the Y-intercept (a), as most of you have probably already noted, to calculate this value one must first calculate the slope. The Y-intercept formula also requires that we calculate the means for the X and Y variables. This is accomplished by simply taking the sum of the X's ($\Sigma X = 24$) and the sum of the Y's ($\Sigma Y = 26$) and separately dividing each of these values by the sample size ($n = 7$); $24/7 = 3.4285 = \bar{x}$, while $26/7 = 3.7142 = \bar{y}$. Next, these values are plugged into the formula to give us $3.7142 - (.8387)(3.4285) = 3.7142 - 2.8754 = .8388 = a$. A summary of all these calculations is found below (intermediate calculations are presented in Table 10.3). Also found (in Figure 10.6) is the previous scattergram with the least-squares regression line plotted from the final derivation of $Y = a + bX$ or $Y = .8388 + (.8387)X$. (Please note that seldom, if ever, are the slope and intercept values this close.)

$$b = \frac{104 - (24)(26)/7}{100 - (24)^2/7} = \frac{104 - 89.1428}{100 - 82.2857} = \frac{14.8572}{17.7143} = \boxed{.8387}$$

$$a = 3.7142 - (.8387)(3.4285) = 3.7142 - 2.8754 = \boxed{.8388}$$

Least-Squares Line ($Y = a + bX$)
or
$Y = .8388 + (.8387)X$
$.8388 + (.8387)(0) = .8388$
$.8388 + (.8387)(1) = 1.6775$
$.8388 + (.8387)(2) = 2.5162$
$.8388 + (.8387)(3) = 3.3549$
$.8388 + (.8387)(4) = 4.1936$
$.8388 + (.8387)(5) = 5.0323$
$.8388 + (.8387)(6) = 5.871$

FIGURE 10.6 Scattergram and Plotted Least-Squares Regression Line

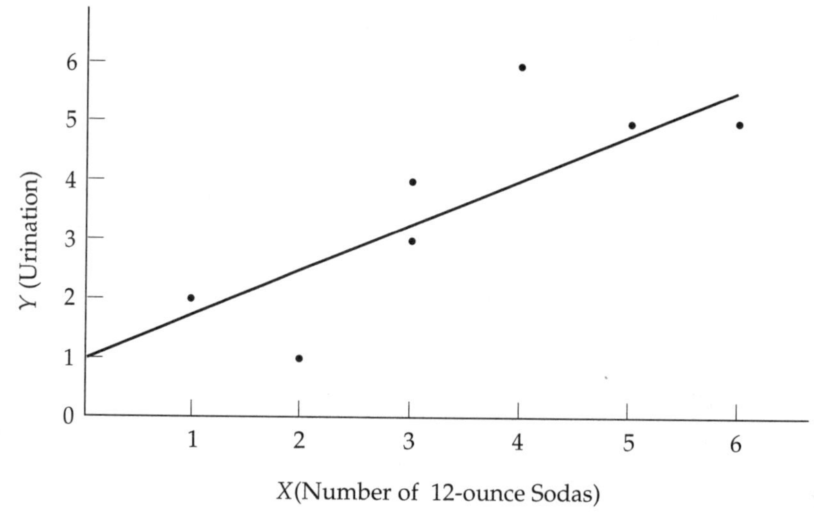

Apparently the consumption of soda does have an effect on the number of times our subjects visit the bathroom. More specifically, with a positive slope we conclude that increased soda consumption apparently leads to increased urination. Moreover, soda and urination are apparently related. The strength of this relationship, however, cannot be assessed using a least-squares regression line. To make this determination we need an additional measure called a correlation coefficient.

Pearson's-*r* Correlation Coefficient

At this point, instead of giving the customary second example, for continuity purposes we now turn to the calculation of what is called a Pearson's-*r* correlation coefficient. (A second example for calculating the least-squares regression line and what follows is given as a summary problem at the end of this chapter.) Whereas the calculation and plotting of the least-squares regression line gives us a clear picture of what sort of relationship exists between two variables (positive or negative), it tells us nothing about the strength of the relationship. This is determined by using the correlation coefficient just mentioned.

Before giving the actual Pearson's-*r* formula, it is helpful to first discuss what it represents. To this end, we reuse the soda/urination data. On Figure 10.7, and exactly as it was previously presented, is the plotted least-squares regression line and the scattergram for this data set. Also imposed onto this figure is a horizontal line representing the mean of the Y variable (3.7142). The amount that each observation deviates from the mean of Y, called the sum of squares, is determined by two factors: regression and error. If the observations are very close to the fitted line, we can infer that most of the deviation is due to regression. Deviation due to regression means the amount of variance in the dependent variable (Y) that is explained —inferred, as caused—by variation in the independent variable (X). Conversely, if the observed values are found predominantly away from the fitted line in no discernible pattern, we then infer that most of the deviation is attributable to error (also called residual variance).

Both of these types of deviation are also graphically presented in Figure 10.7. So, with individual 2, who consumed two 12-ounce sodas and urinated once, we see that the amount of variation to the least-squares line is illustrated as due to regression while the amount it goes beyond the observed value is illustrated as a result of error. Stating this slightly differently, individuals who are given two 12-ounces sodas are expected to urinate 2.5162 times; however, our research subject who drank two ounces of soda only urinated once. As such and in simplistic terms, the difference between the expected and the observed value, $2.5162 - 1 = 1.5162$, is the amount of variance due to error. The distance from the mean of Y to the expected value is, conversely, the amount of variance due to regression: $3.7142 - 2.5162 = 1.198$. In other words, this is the amount of variation in the dependent variable that is seen as being caused by the independent variable. Remember, however, that although we are inferring causality, all we can demonstrate absolutely is the amount that two variables are correlated—associated—together.

A more precise way of measuring this is using the Pearson's-*r* correlation coefficient formula. The resultant answer (squared, as we will see) from this formula tells us

FIGURE 10.7 Estimated Amount of Deviation Due to Error and Regression

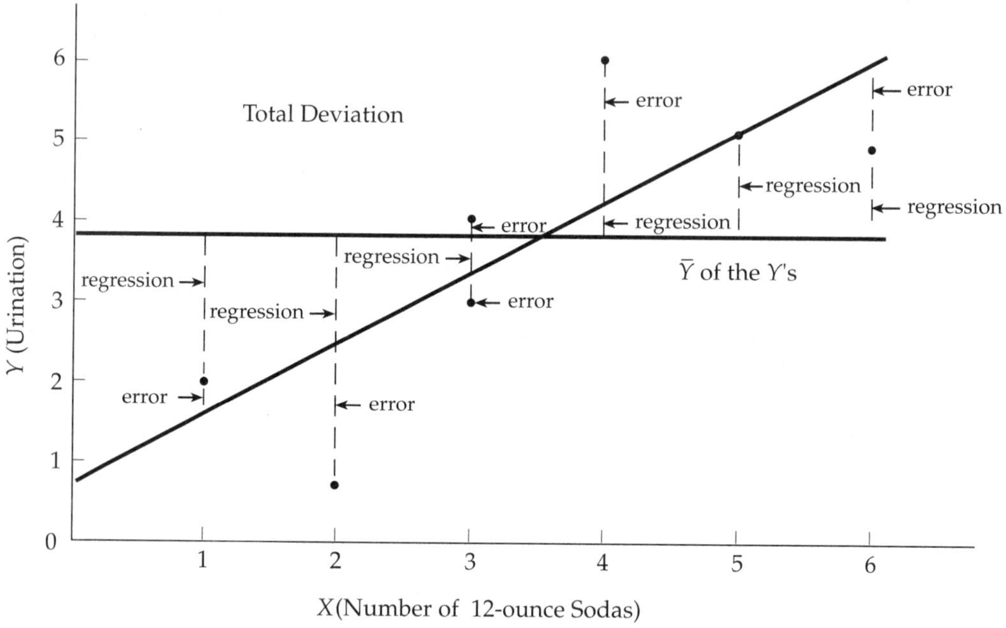

exactly how much of the variation is attributed to regression versus error. The answer this formula gives us is on a scale of 0 to 1 (actually -1 to $+1$; in cases where a negative coefficient is given, once again, this informs us that we have a negative inverse relationship). The closer the final answer is to the value of 1, the more the two variables are associated together and the greater the amount of variability in the dependent variable is inferred as due to regression. Conversely, the closer the final answer is to 0, the less the variables are associated together; as such, more of the variability is inferred as due to error. What the actual formula does is to determine the total amount of deviation and then divide this value by the amount of deviation due to regression.

The actual Pearson's-*r* formula is found below and looks very similar to the previously discussed slope (*b*) formula except that now we must take the square root of the previous denominator of *X* and multiply it by the square root of the *Y* calculation. What it mathematically does is determine the amount of variance similar to two variables, called covariance (the numerator portion of the formula), and divides this value by the average amount of variance for each variable (the denominator portion of the formula). Since we have previously completed most of the calculations, we have simply presented the actual calculations for this problem below the formula (Table 10.4 is also offered to assist in this presentation). (Only one note is warranted: As is done below, the resultant square-root values in the denominator are multiplied together.)

TABLE 10.4 Intermediate Steps for Soda/ Urination Data (Pearson's-*r* Formula)

	X	Y	X^2	Y^2	XY
1	1	2	1	4	2
2	2	1	4	1	2
3	3	3	9	9	9
4	3	4	9	16	12
5	4	6	16	36	24
6	5	5	25	25	25
7	6	5	36	25	30
	24	26	100	116	104

Pearson's-*r* Formula

$$\frac{\Sigma XY - (\Sigma X)(\Sigma Y)/n}{\sqrt{\Sigma X^2 - (\Sigma X)^2/n}\ \sqrt{\Sigma Y^2 - (\Sigma Y)^2/n}}$$

Calculations

$$r = \frac{104 - (24)(26)/7}{\sqrt{100 - (24)^2/7}\ \sqrt{116 - (26)^2/7}} = \frac{14.8572}{\sqrt{17.7143}\ \sqrt{19.4286}}$$

$$= \frac{14.8572}{(4.2088)(4.4077)} = \frac{14.8572}{18.5511} = \boxed{.8008}$$

With a Pearson's-*r* (correlation coefficient) of .8008, the variables of soda and urination are strongly related (associated). Further (as we have already discovered when calculating the slope), since the answer is positive, the relationship between the two variables is a strong, positive one. While the answer (.8008) tells us that the two variables are strongly associated together, it does not tell us the amount of variance that is due to regression versus error. To answer this question we use what is called the *coefficient of determination*: the total variation of the Y variable that is a function of variation in the X variable. This is simply derived by squaring the value of the obtained correlation coefficient. This is done below.

$$r = .8008 \qquad r^2 = \boxed{.6412} = \text{Coefficient of Determination}$$

From this final answer, we can now infer that 64.12% of the variance in the dependent variable is due to regression. Conversely, we can also infer that 35.88% (1 − .6412 = .3588) of the variance in the dependent variable is due to error: unexplained variance. Thus, we conclude (although not absolutely) that 64.12% of the number of times our seven respondents urinated is attributable to the number of sodas they consumed while 35.88% of the number of times they urinated is error or unexplained variance. Obviously, we can also conclude that drinking soda causes one to urinate, and the more soda one drinks the greater the number of times they will urinate.

TESTING FOR SIGNIFICANCE

While we now have a clear picture of what the relationship between soda and urination looks like and how strong it is, we do not know if the relationship is statistically significant. To do this, we use a variant form of *t* test found below. Since we have already determined each of the symbols within the formula, its calculation should be quite easy for you. Once again, however, since this is our first time through a new formula, we review each symbol by verbally doing the required calculations.

$$t \text{ test for Pearson's } r = r\sqrt{\frac{n-2}{1-r^2}}$$

As the formula asks, we first take the sample size and subtract 2 from it; $7 - 2 = 5$. Then we take the value of 1 and subtract *r* squared from it; $1 - .6412 = .3588$. Next, the value of 5 is divided by .3588 ($5/.3588 = 13.9353$); we then take the square root of this value to get 3.733. This value of 3.733 is multiplied by *r* (.8008), which gives us the final answer—a *t* obtained of 2.9893.

As with previous *t* tests, the obtained value is now compared to a critical value. While the *t*-critical value is discerned using the familiar alpha level, the determination of degrees of freedom (df), however, is slightly different with this type of *t* test. Since we are dealing with two variables, we use $n - 2 = \text{df}$ instead of the previous $n - 1$. Thus for the present example we take $7 - 2$, giving us 5 df. Setting the alpha level at .05, the corresponding *t*-critical value is 2.571. With the obtained value of 2.9893 being larger than the critical value of 2.571, we conclude that the more sodas one drinks the more likely one is to urinate, and this relationship is statistically significant. All of the steps for calculating this type of *t* test applied to the soda/urination example are mathematically summarized below.

<u>*t* test for Soda/Urination Data</u>

$$r\sqrt{\frac{n-2}{1-r^2}} = .8008\sqrt{\frac{7-2}{1-.6412}} = .8008\sqrt{\frac{5}{.3588}}$$

$$= .8008\sqrt{13.9353} = (.8008)(3.733) = \boxed{2.9893} = t \text{ obtained}$$

(alpha = .05, 5 df = *t* critical = 2.571)

SUMMARY EXAMPLE

To give a second set of examples for all of the formulas that have been discussed in the chapter, we offer the following summary example based upon Cartoon 10.1. This cartoon is also used for the Practice Exercises.

Reflecting this cartoon, we offer the following data set (see Table 10.5) representing a 12-month period of January 1 through December 24. The independent variable (*X*) is the number of complaints received per month about Santa's herd of reindeer. The dependent variable (*Y*) is the current size of Santa's reindeer herd at the end of each month.

The data also reflect a negative relationship culminating in the month of

CARTOON 10.1

THE FAR SIDE By GARY LARSON

And I've only one thing to say about all these complaints I've been hearing about . . . venison!

TABLE 10.5 Complaints and Herd Size Data

Month	*Number of Complaints Received Per Month* (X)	*Current Size of Herd* (Y)
January	2	25
February	1	24
March	3	24
April	4	23
May	3	22
June	5	20
July	4	20
August	7	18
September	8	17
October	10	14
November	11	12
December	14	9

FIGURE 10.8 Scattergram for Venison Data

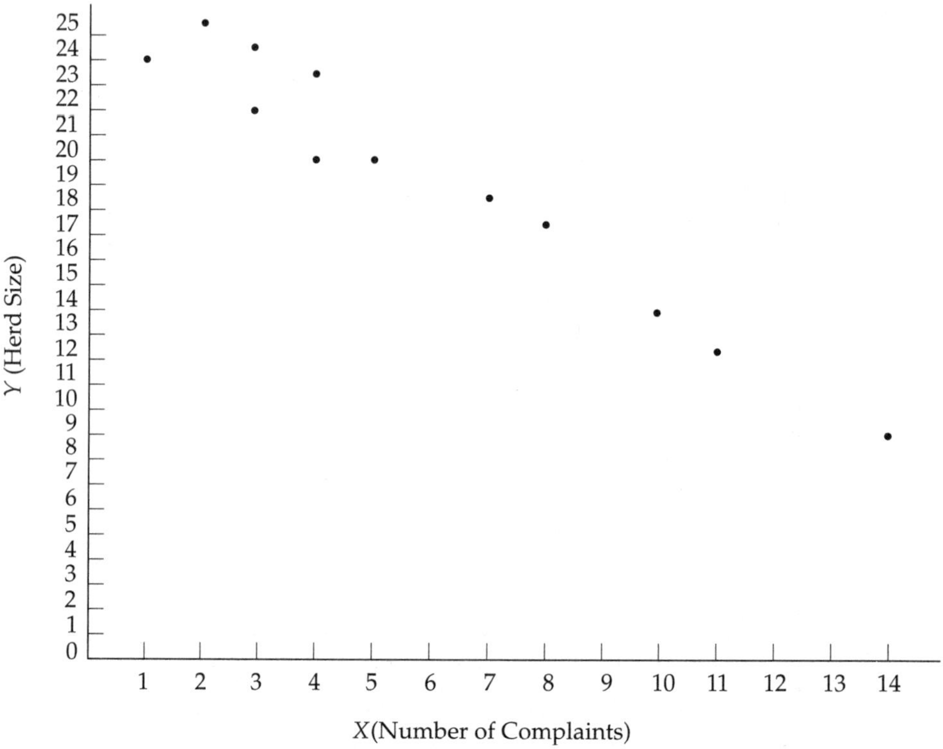

December with the traditional nine reindeer left to pull Santa's sled. In other words, each month leading up to Christmas sees more complaints with fewer reindeer left in Santa's herd. (This relationship can also be conceptualized as a positive one; that is, more complaints lead to more reindeer being turned into venison. This example, however, is presented as a negative, inverse relationship to give you an idea of some of the minor differences in plotting and calculation.) Using the previously explored formulas, we now demonstrate mathematically what type of relationship exists, what its strength is, and see if it is statistically significant.

As previously done, we first plot the data in the form of a scattergram. Once again, the independent variable (number of complaints) is the horizontal axis, while the dependent variable (size of herd) is the vertical axis. These data are plotted in Figure 10.8.

Having constructed our scattergram, we now turn to the calculation of our least-squares regression line; the slope (b) and the intercept (a). Found on Table 10.6 is the data and the intermediate operations that are used in the formulas. Presented below are the slope and intercept formulas to aid in this presentation, and the final calculations.

Formulas for Slope (b) and Intercept (a)

$$b = \frac{\Sigma XY - (\Sigma X)(\Sigma Y)/n}{\Sigma X^2 - (\Sigma X)^2/n} \qquad a = \bar{y} - b\bar{x}$$

$$= \frac{1144 - (72)(228)/12}{610 - (72)^2/12} \qquad = 19 - (-1.2584)(6)$$

$$= \frac{1144 - 1368}{610 - 432} \qquad = 19 - (-7.5504)$$

$$= \frac{-224}{178} = \boxed{-1.2584} \qquad = \boxed{26.5504}$$

TABLE 10.6 Intermediate Calculations for Complaints and Herd Size

Month	*Number of Complaints Received Per Month* (X)	*Current Size of Herd* (Y)	XY	X^2	Y^2
January	2	25	50	4	625
February	1	24	24	1	576
March	3	24	72	9	576
April	4	23	92	16	529
May	3	22	66	9	484
June	5	20	100	25	400
July	4	20	80	16	400
August	7	18	126	49	324
September	8	17	136	64	289
October	10	14	140	100	196
November	11	12	132	121	144
December	14	9	126	196	81
Sums	72	228	1144	610	4624

With a slope of −1.2584, we confirm what is already known; this is a negative, inverse relationship. Further, when a negative slope value is used to calculate the least-squares line (below), we actually make this value minus after it has been multiplied by each value of X. Moreover, to calculate the intercept for this example, when a negative value is subtracted we actually end up adding it—minus a minus means we add the value; 19 − (−7.5504) = 26.5504. Given this information, we have plotted the least-squares regression line over the previously constructed scattergram (see Figure 10.9).

Least-Squares Line = $Y = a + bX$	
$Y = 26.8388 + (-1.2584)X$	(continued)
26.5504 + (−1.2584) (0) = 26.5504	26.5504 + (−1.2584) (8) = 16.4832
26.5504 + (−1.2584) (1) = 25.292	26.5504 + (−1.2584) (9) = 15.2248
26.5504 + (−1.2584) (2) = 24.0336	26.5504 + (−1.2584) (10) = 13.9664
26.5504 + (−1.2584) (3) = 22.7752	26.5504 + (−1.2584) (11) = 12.708
26.5504 + (−1.2584) (4) = 21.5168	26.5504 + (−1.2584) (12) = 11.4496
26.5504 + (−1.2584) (5) = 20.2584	26.5504 + (−1.2584) (13) = 10.1912
26.5504 + (−1.2584) (6) = 19.00	26.5504 + (−1.2584) (14) = 8.9328
26.5504 + (−1.2584) (7) = 17.7416	

With the observed values falling very close to the plotted regression line, as the above graphically represents, we obviously have a very strong linear relationship that is inverse. To mathematically demonstrate the strength of the relationship, once again we use the Pearson's-*r* formula. Both the formula and the actual calculations for this data set are presented below. The coefficient of determination value, *r* squared, is also found below.

Pearson's-*r* Formula

$$\frac{XY - (\Sigma X)(\Sigma Y)/n}{\sqrt{\Sigma X^2 - (\Sigma X)^2/n}\ \sqrt{\Sigma Y^2 - (\Sigma Y)^2/n}}$$

Calculations

$$r = \frac{1144 - (72)(228)/12}{\sqrt{610 - (72)^2/12}\ \sqrt{4624 - (228)^2/12}} = \frac{-224}{\sqrt{178}\ \sqrt{292}}$$

$$= \frac{-224}{(13.3416)(17.088)} = \frac{-224}{227.9812} = \boxed{-.9825}$$

$$r = -.9825; \quad r^2 = \boxed{.9653} = \text{Coefficient of Determination}$$

With the obtained *r* value being negative, we are once more told that this is a negative, inverse relationship. Once the *r* value of −.9825 is squared, the resultant coefficient-of-determination value of .9653 is quite high on a scale of 0 to absolute 1. Moreover, this value allows us to infer that 96.53% of Santa's herd size is explained by

FIGURE 10.9 Plotted Least-Squares Regression Line for Venison Data

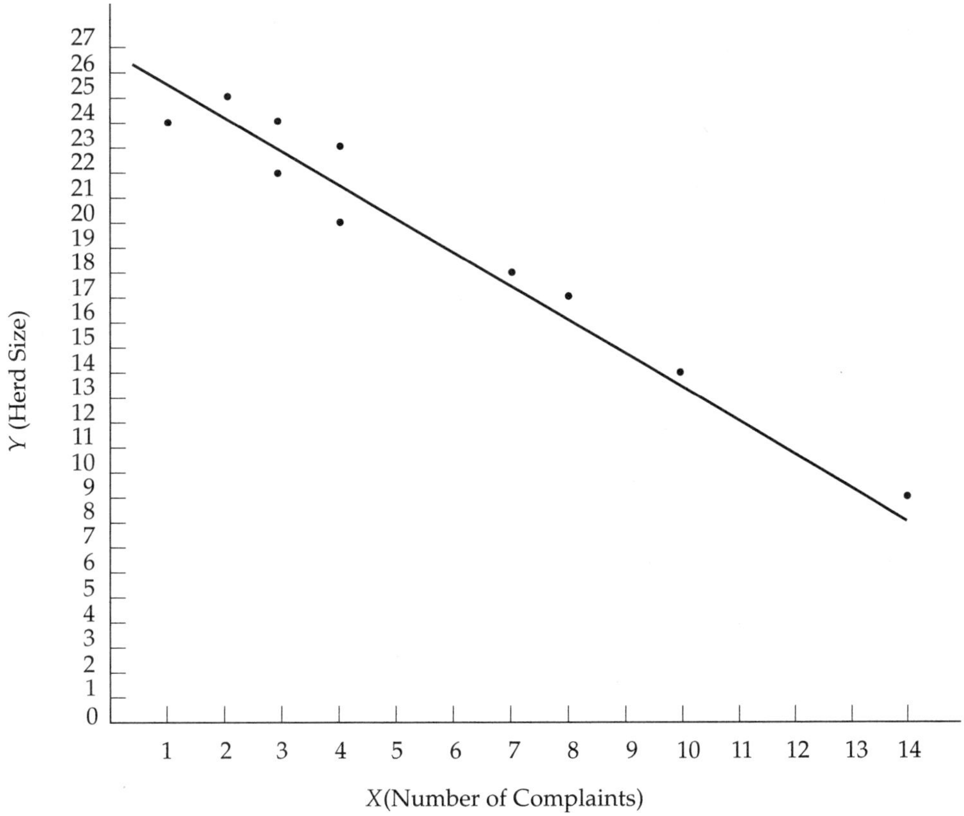

the number of complaints received in a given month; only 3.47% is due to unexplained variance—error. In other words, complaints have a deadly effect on Santa's reindeer herd. While this is obviously significant, as presented below, we have still to complete the t test for this data set:

$$t \text{ Test for Pearson's } r = r\sqrt{\frac{n-2}{1-r^2}}$$

<u>t Test For Complaints/Herd Size Data</u>

$$-.9825\sqrt{\frac{12-2}{1-.9653}} = -.9825\sqrt{\frac{10}{.0347}}$$

$$= -.9825\sqrt{288.1844} = (-.9825)(16.9759) = \boxed{-16.6788} = t \text{ obtained}$$

(alpha = .001, 10 df = t critical = 4.587)

In sum, we conclude that an increase of complaints leads to a statistically significant reduction in Santa's herd size. This may explain how Santa gets his reindeer to fly. That is, if you are a reindeer and lucky enough to have made it to Christmas, given the alternative you would probably fly too.

Box 13

While all the examples in this chapter have used interval-ratio variable relationships, there are often situations where ordinal and nominal variables must be used. In these circumstances we are required to use different measures of association. Before outlining these other measures, note that although Pearson r has the explicit assumption that interval-ratio data are used in its calculation, this assumption is often violated. While numerous researchers and statisticians have offered various explanations for violating this assumption of using interval-ratio data (and it has become fairly commonplace to do so), we still feel it important that you understand that there are other statistical techniques developed to deal with this problem. Since it is beyond the scope of an introductory statistics textbook to explore these alternative measures of association, we have simply outlined what they are in Table 10.7.

TABLE 10.7 Types of Correlation Coefficient for Relationships Among Nominal, Ordinal, and Interval-Ratio Variables

		Independent Variable		
		Interval-Ratio	*Ordinal*	*Nominal*
Dependent Variable	Interval-Ratio	Pearson r	Biserial r	Point Biserial r
	Ordinal	Biserial r	Spearman's Rho	Phi
	Nominal	Point Biserial r	Phi	Phi

Chapter Summary and Conclusions

This chapter introduced a new way of conceptualizing bivariate relationships. First we looked at such relationships in terms of a graphic illustration, a scattergram. Next we built upon the scattergram to demonstrate that, using the least-squares regression formula, a straight line can be determined and plotted to represent the entire data set. While this information gave us a general idea of whether the relationship was positive or negative and what it looked like, we had to explore additional formulas to determine the strength of the relationship and whether it is statistically significant. To this end, we used the Pearson's-r formula, a coefficient of determination (r squared), and a newly introduced t test. The chapter ended by noting that, dependent upon what type of variable is being used, there are several other measures of association that can be utilized.

Key Terms To Remember

Independent Variables	Scattergram
Dependent Variables	Least-Squares Regression Line
Notion of Causality	Pearson's-r Correlation Coefficient
Slope	Coefficient of Determination
Y-Intercept	Pearson's-r/t test

Practice Exercises

1. The data set in Table 10.8 is in reference to Cartoon 10.1. As labeled, the independent variable represents the previously given number of complaints per month for Santa's reindeer. The dependent variable, however, represents the number of reindeer turned into venison.

TABLE 10.8 Complaints and Venison Data

Month	*Number of Complaints Received Per Month* (X)	*Number Turned Into Venison*(Y)
January	2	1
February	1	2
March	3	3
April	4	2
May	3	5
June	5	8
July	4	7
August	7	10
September	8	9
October	10	13
November	11	16
December	14	13

a. Construct a scattergram with this data set. What type of relationship appears to exist between complaints and reindeer becoming venison?
b. Calculate the slope (a) and intercept (b) for this data set.
c. Plug the calculated slope and intercept values into the least-squares regression line ($Y = a + bX$), and then plot it over the previously constructed scattergram.
d. Calculate the Pearson's-r correlation coefficient for this data set.
e. Using the appropriate t test, determine if the relationship between complaints and number of reindeer becoming venison is statistically significant.
f. Having made all of these calculations, what would one conclude about the relationship between complaints and number of reindeer becoming venison?

2. Once again, in reference to this cartoon, we are given the data set in Table 10.9. The independent variable is the age of the reindeer while the dependent variable represents the number of complaints received in a year's time. This data set obviously assumes that these 10 reindeer did not become venison in the (year period) of measurement.

a. Construct a scattergram with this data set. What type of relationship appears to exist between age and number of complaints received in a year?
b. Calculate the slope (a) and intercept (b) for this data set.
c. Plug the calculated slope and intercept values into the least-squares regression line ($Y = a + bX$), and then plot it over the previously constructed scattergram.
d. Calculate the Pearson's-r correlation coefficient for this data set.
e. Using the appropriate t test, determine if the relationship between complaints and number of reindeer becoming venison is statistically significant.
f. Having made all of these calculations, what can you conclude about the relationship between age and the number of complaints received?

TABLE 10.9 Age and Complaints Data

Reindeer's Name	*Age of Reindeer*(X)	*Number of Complaints Received for the Year*(Y)
Rudolph	1	20
Donner Jr.	1	15
Dasher II	3	17
Blixen	3	14
Dancer	4	15
Prancer	5	11
Comet	5	8
Cupid	5	10
Donner Sr.	7	9
Dasher III	10	8

Chapter 11

One-Way Analysis of Variance (ANOVA)

Chapter 9 explored various techniques of hypothesis testing where group means were compared to determine if there was a significant difference between them. While these statistical tests were concerned with just two group means, this chapter explores situations where three or more group means are compared. The specific statistical technique to accomplish this is called *analysis of variance*, hereafter referred to as ANOVA (the actual ANOVA procedure is also referred to as an *F* test).

More specifically, this chapter first outlines some of the advantages and theoretical assumptions that explain why ANOVA is often utilized as a statistical technique. Then, using an 18-step procedure, the actual calculations for completing an ANOVA are presented. When statistical significance is determined using ANOVA, subsequent statistical tests are often undertaken. One of these tests, called an *LSD t test*, is also explored in terms of its calculation and what it represents. The chapter ends with some simple examples of *ANOVA Summary Tables* and how to interpret them.

Advantages and Theoretical Assumptions of ANOVA

Like many statistical tests, ANOVA originated from studies in agriculture. Researchers in the agricultural sciences often study things such as the effects of different levels of irrigation on land plots with varied levels and types of fertilizers. In investigations such as this, the researcher is often faced with situations where dozens of group means must be compared. This translates into literally hundreds of potential *t* tests; anything over 11 groups correspondingly results in a minimum of 100 possible *t* tests.

The way we determine the number of potential *t* tests that can be performed is to use the following formula (*J represents the number of groups*). If we wanted to know the number of possible t tests that can be performed for seven group means, the value of seven is simply plugged into the formula to find that there are 21 possible t tests. These calculations are found below with the actual formula.

$$\frac{J(J-1)}{2} \quad \text{or} \quad \frac{7(7-1)}{2} = \frac{42}{2} = \boxed{21}$$

While the calculation of 21 t tests is no more difficult than the amount of time it takes to complete them, undertaking a large number of t tests, such as 21, does present a theoretical dilemma. That is, if we had 20 possible t tests (now referred to as 20 to simplify the discussion) and the alpha level was set at .05, how many t values would we expect to find significant when in fact they are not significant? What we are actually referring to here is the probability of committing a *Type I Error*: rejecting the null hypothesis when in fact it is true. (At this point, to refresh your memory, you may want to refer back to Chapter 9 for a more comprehensive discussion of Type I Errors.) With 20 t tests, alpha level set at .05, due to potential sampling error, we expect one of these t tests to be found significant when in fact it is not: once again, a Type I Error.

Because of this potential predicament, when researchers are undertaking analyses that involve the calculation of numerous potential t tests they need an additional test to guard against committing Type I Errors. This is why the statistical procedure of ANOVA was created. (Although this technique was created to deal with situations of three or more group means, it can also be used with just two group means.) ANOVA enables researchers to determine statistically if additional tests (including t tests and an array of other tests) are warranted. Quite simply, if one obtains a *significant F* value using ANOVA, then *subsequent statistical tests* are typically undertaken. Conversely, if an F value that is not significant is obtained, then subsequent tests should not be undertaken. This becomes clearer when we proceed to the actual calculations.

The actual ANOVA statistical test informs us if there is more variance *between* the groups than that which is found *within* all of the groups. More specifically, given any set of numbers we almost always find variability. One way the variability within a group is measured and represented is with a variance value (the actual calculations for this formula were presented in Chapter 5). The variability between groups is measured in a similar manner. What the ANOVA formula represents, as a final answer, is the amount of variability between the groups divided by the amount of variability within the groups. In other words, the *obtained F value* is a *ratio value*.

As demonstrated shortly, if there is measurably more variability between the groups than within them, the obtained-ratio value (F) will be found to be statistically significant. Conversely, if the amount of variability between the groups is close to or less than the amount of variability found within the groups, the obtained F value will be close to or less than one; such an answer would be statistically insignificant.

Applying the *null* and *research hypotheses* to ANOVA, we find that instead of statements of one mean being or not being statistically equal, these potential relationships are stated in slightly different terms. When three or more group means are compared, the null hypothesis simply states that none of the means is significantly different; all of them are equal. Conversely, the research hypothesis holds that one or more of the means are statistically different, not equal. If statistical significance is found using ANOVA, it simply tell us that one or more of the means are not equal; however, it does not specify which of the means are not equal. Subsequent statistical tests, such as the LSD *t* test, are used to determine which of the groups means are actually statistically different. Both the null and research hypotheses for an ANOVA are mathematically represented below.

Null Hypothesis	Research Hypothesis
H_0: $\mu_1 = \mu_2 = \mu_3$	H_A : One or more group means are not equal.

One final preliminary observation is warranted. As with previous tests, once an obtained value is derived, this value has to be compared to a critical table value to determine if there is statistical significance. Since this is an *F* test, not surprisingly, a new set of *F*-critical values, found in Appendix 3, are used to make this determination. For simplicity purposes, the table values found in Appendix 3 list *F-critical values* at two alpha levels: .05 and .01. Here, like previous critical values, we are making a statement of probability of the likelihood of a Type I Error having been committed. In other words, even when using ANOVA, due to sampling error the potential still exists that a Type I Error might be committed. This technique does, however, greatly diminish the probability that a Type I Error is committed with subsequent tests; once again, these tests cannot be performed unless a significant *F* value is obtained.

ONE-WAY ANALYSIS OF VARIANCE (ANOVA)

We are now ready to turn to the actual calculations used to complete an ANOVA. To assist in this discussion we use Cartoon 11.1 found below. Let's say our current research interests are the effects of different types of liquids poured "off" a duck's back. We randomly select three groups of 10 ducks, 30 ducks total, for our experiment (for this technique to be used in a theoretically appropriate manner, the selection of the research subjects—ducks—and their placement into each of the groups should be randomly done). Group 1 receives the treatment of one ounce of acid poured "off" their backs. Group 2 receives the treatment of one ounce of vinegar poured "off" their backs. Group 3 is seen as our control group and no treatment is given; nothing is poured "off" their backs. (Although a control group is often utilized when undertaking an ANOVA, it need not be.) Each group of ducks is then observed for 24 hours to determine the number of feathers each of them loses or gains during this period. The findings for this experiment are reported in Table 11.1.

Since some of the symbols found in this table are being presented for the first time, we will review all of them. To begin with, the table reports the findings for our three groups of ducks in terms of the number of feathers lost or gained at the end of the 24 - hour period. This is indicated by the very top column headings. Under these column headings are the sub-headings of *s*; *X*.1, *X*.2, or *X*.3; and *X*.1, *X*.2, or *X*.3 squared. The symbol *s* simply represents research subjects (ducks) for a given group (e.g., *X*31 or *X*93). The column headings *X*.1, *X*.2 , and *X*.3 indicate which group the observation is taken from. Thus, for instance, research subject *X*42 lost one feather as a result of par-

TABLE 11.1 Liquids off a Duck's Back Data

One-way Analysis of Variance for Feathers Lost (or Gained)

	Group 1			*Group 2*			*Group 3*	
s	*X.1*	$X.1^2$	*s*	*X.2*	$X.2^2$	*s*	*X.3*	$X.3^2$
X11	−8		X12	+2		X13	−3	
X21	−9		X22	−6		X23	+2	
X31	−10		X32	−4		X33	−3	
X41	−8		X42	−1		X43	+2	
X51	−11		X52	−1		X53	+4	
X61	−15		X62	+4		X63	−1	
X71	−16		X72	−3		X73	−3	
X81	−7		X82	−2		X83	−2	
X91	−12		X92	−4		X93	+2	
X101	−14		X102	−5		X103	−3	
ΣX.1			ΣX.2			ΣX.3		
$\Sigma X.1^2$			$\Sigma X.2^2$			$\Sigma X.3^2$		
$(n_1 = 10)$			$(n_2 = 10)$			$(n_3 = 10)$		

Group 1: One ounce of acid poured "off" a duck's back.
Group 2: One ounce of vinegar poured "off" a duck's back.
Group 3: Control group; no treatment given.

ticipating in this experiment while subject X31 lost 10 feathers. The column heading X.1, X.2, and X.3 squared are used for calculation purposes and are simply any observed value squared. Finally, at the *bottom* of each group's *X* and *X* squared columns are spaces reserved for their totals: (1) the upper row represents the sum of all the observed values for the given group and (2) the lower row of the matrix represents the sum of all the observed values squared for the given group.

As already noted, this text treats the calculations for ANOVA as an 18-step procedure. While each of the steps is individually completed below, for clarification purposes and as a suggested way for you to do your own ANOVA, answers for all 18 calculations are also separately summarized. Further, as demonstrated shortly, the last eight calculations are also placed on an ANOVA summary table.

The first step asks us to find the sum of all the *X* values. This is accomplished by summing the observed values separately for each of the groups, and then adding together the groups' summed values. The *i* and *j* subscripts found in the formula simply denote, from beginning to end, the sum of all observed values. With the present or similar examples (where both negative and positive values are summed), one must be very careful to differentiate between these two types of values; otherwise, as you might guess, any subsequent answers are incorrect. These operations are completed in a revised table of raw scores (Table 11.2) and are mathematically represented and completed as follows.

CARTOON 11.1

Step 1: The sum of $\Sigma X_{ij} = \Sigma X1 + \Sigma X2 + \cdots + \Sigma X_j$

For this example: $\Sigma X1 + \Sigma X2 + \Sigma X3$

$$= -110 + -20 + -5 = \boxed{-135}$$

Step 2 also requires calculations to be made in the table of raw values. To this end, we are required to first square each observation of X (answers are found in column 2 for each group), then separately sum these values for each group, and finally add together the three group sums into one final answer. The completed operations for this step are found below and in the revised table (Table 11.2) of raw values.

Step 2: The sum of $\Sigma X_{ij} = \Sigma X_1^2 + \Sigma X_2^2 + \Sigma X_3^2 + \cdots + \Sigma X_j^2$

For this example: $\Sigma X1_1^2 + \Sigma X2_2^2 + \Sigma X3_3^2$

$$= 1300 + 128 + 69 = \boxed{1497}$$

The next operation is not required in terms of obtaining the final answer; nevertheless, Step 3 does give us an interesting, alternative look at our data set. This step asks us to sum all the X values, as we did in Step 1, and then divide this value by the total number of observations for all groups. We represent the later symbol as N, which in this case is 30. [Even though we are actually dealing with a sample, we use an upper-case N instead of a lower-case n to aid in subsequent calculations. A proceeding ANOVA step, as we will see, uses lower-case n's to represent individual group sizes

TABLE 11.2 Intermediate Steps for Liquids off a Duck's Back Data

One-Way Analysis of Variance for Feathers Lost (or Gained). Revised Table With Squared and Summed Table Values Included.

	Group 1			*Group 2*			*Group 3*	
s	*X.1*	*X.1²*	*s*	*X.2*	*X.2²*	*s*	*X.3*	*X.3²*
X11	−8	64	X12	+2	4	X13	−3	9
X21	−9	81	X22	−6	36	X23	+2	4
X31	−10	100	X32	−4	16	X33	−3	9
X41	−8	64	X42	−1	1	X43	+2	4
X51	−11	121	X52	−1	1	X53	+4	16
X61	−15	225	X62	+4	16	X63	−1	1
X71	−16	256	X72	−3	9	X73	−3	9
X81	−7	49	X82	−2	4	X83	−2	4
X91	−12	144	X92	−4	16	X93	+2	4
X101	−14	196	X102	−5	25	X103	−3	9
ΣX.1	−110		ΣX.2	−20		ΣX.3	−5	
ΣX.1²		1300	ΣX.2²		128	ΣX.3²		69

($n_1 = 10$) ($n_2 = 10$) ($n_3 = 10$)

Group 1: One ounce of acid poured "off" a duck's back.
Group 2: One ounce of vinegar poured "off" a duck's back.
Group 3: Control group; no treatment given.

(e.g., the n for group 1 is 10).] What the final answer for Step 3 actually tells us is that the grand (total) mean for all observations is −4.5. In other words, our 30 ducks lost an average of 4.5 feathers.

One further observation is warranted at this point; Steps 1 and 3 are the only ones where negative answers can be obtained. Otherwise, every other ANOVA step involves the squaring of values; therefore, all other resultant answers, including the final answer, are always positive values.

Step 3: $$\frac{\Sigma X}{N} = \frac{(\text{Step 1})}{N} = \frac{-135}{30} = -4.500$$

We are now ready to make calculations that represent the total amount of variance found in each of the groups. This is also referred to as the total square sum of scores (divided by n). To accomplished this, the sum of X for each group is separately squared, and then the resultant answer is divided by the group size. These calculations need not be done separately when each of the group sizes is the same; this is true for the case at hand ($n_1 = 10, n_2 = 10$, and $n_3 = 10$). Since, however, most data sets do not entail equal group sizes, we have done each of the three calculations separately. These calculations (listed as Steps 4, 5, and 6) are found below.

Calculations for Steps 4, 5, and 6

Obtain the $\Sigma(\Sigma X_{ij})^2 / n_{ij}$ for each available group.

Step 4: $$\frac{(\Sigma X.1)^2}{n_1} = \frac{(-110)^2}{10} = \frac{12100}{10} = \boxed{1210}$$

Step 5: $$\frac{(\Sigma X.2)^2}{n_2} = \frac{(-20)^2}{10} = \frac{400}{10} = \boxed{40}$$

Step 6: $$\frac{(\Sigma X.3)^2}{n_3} = \frac{(-5)^2}{10} = \frac{25}{10} = \boxed{2.5}$$

The next step is really what the formula prior to Steps 4–6 asked for: the total square sum of scores divided by n. In simpler terms, to complete Step 7, add Steps 4, 5, and 6 together.

Step 7: $$\frac{\Sigma(\Sigma X_{ij})^2}{n_{ij}} = \text{Step 4} + \text{Step 5} + \text{Step 6}$$

$$= 1210 + 40 + 2.5 = \boxed{1252.5}$$

Step 8 is an intermediate operation and requires that we sum all the observed X's, square this value, and then divide it by the total number of subjects (N). In other words, take Step 1, square it, and then divide this answer by N. This step is mathematically presented as follows.

Step 8: $$\frac{(\Sigma X_{ij})^2}{N_{ij}} = \frac{(\text{Step 1})^2}{N_{ij}} = \frac{(-135)^2}{30} = \boxed{607.5}$$

At this point, all of the answers for subsequent operations are further summarized on what is called an *ANOVA Summary Table* (Table 11.3). This table, along with the summarized 18 steps (Table 11.4), is found at the end of all of the calculations. In turn, each of the new symbols in Table 11.3 is explained as its corresponding calculations are made. Not only is this table an efficient way to summarize meaningful information about our data set and the mathematical computations we have made, but some of the values (other than the final answer) are used in subsequent statistical tests.

Step 9, whose answer is the first to be placed in the ANOVA Summary Table, tells us how much variability between the groups is present; this is referred to as the sum of squares (SS) between groups. To obtain this value, we simply subtract Step 8 from Step 7, and the value of 645 is obtained. This value, 645, is placed in the table under the column heading SS and the row heading Between.

Step 9: The sum of squares between groups = Step 7 − Step 8

= 1252.5 − 607.5 = **645**

Similarly, Step 10 gives us the answer that is placed in the table for the SS within groups. This value is obtained by taking the sum of all individually squared values (Step 2) and subtracting the total square sum of scores value (Step 7). The obtained value of 244.5, reported below, is placed in the table under the column heading SS and the row heading Within.

Step 10: The sum of squares within groups = Step 2 − Step 7

= 1497 − 1252.5 = **244.5**

Although the next step is not required for subsequent calculations, we have included Step 11 as a way to check if previous answers are correct. This step, which gives us the value for the TOTAL sum of squares, is frequently reported in articles and papers that have used an ANOVA statistical technique. Further, for calculation purposes we are required to know this value for some of the subsequent ways ANOVA Summary Tables are explored. As reported below, the answer is obtained by taking the sum of all squared values (Step 2) and subtracting Step 8 from it. This gives us the answer of 889.5, which is placed in the table under the column heading SS and the row heading of TOTAL.

To check our previous calculations, Step 9's answer is compared to the separately derived answer of the SS within added to the SS between (645 + 244.5). In the case at hand, both values equal 889.5. If, however, these values are not equal, a mistake has been made somewhere in the previous calculations and we need to go back and correct our error. Although the potential still exists that these two answers will be equal even when errors have been made in previous calculations, this is, nevertheless, a fairly easy way to check for possible mistakes.

Step 11: The TOTAL sum of squares = Step 2 − Step 8

= 1497 − 607.5 = **889.5**

As you become more proficient at completing ANOVA Summary Tables, all of the remaining steps should come somewhat automatically. For our first few times through these steps, however, we explicitly list each of the required steps. Step 12 represents the number of groups we are dealing with in our analysis. Since this is a sample, one of these groups (conceptualized as a degree of freedom) is fixed. This is done mathematically by taking the total number of groups (represented by the symbol J) and subtracting the value of one from it. In this case, the answer is 2; $J - 1$ or $3 - 1 = 2$. Thus, as reported below and on the ANOVA Summary Table, there are 2 df for this example.

Step 12: The between degrees of freedom $= J - 1 = 3 - 1 = \boxed{2 \text{ df}}$

Next we determine the within degrees of freedom. Using an ANOVA procedure, in a sense, each group is treated as a sub-sample: a discrete, separate set of individuals. As such, one degree of freedom from each individual group is fixed. Mathematically, as found below and reported in the ANOVA table, this is obtained by taking the value of N and subtracting the number of groups (J) from it.

Step 13: The within degrees of freedom $= N - J = 30 - 3 = \boxed{27 \text{ df}}$

Although the next step is another operation not required for any subsequent calculations, it is still a quick way to check if the calculations for Steps 12 and 13 are correct. Quite simply, Step 14 represents the TOTAL degrees of freedom and is obtained by simply subtracting the value of one from N (as found below).

Step 14: The TOTAL degrees of freedom $= N - 1 = 30 - 1 = \boxed{29 \text{ df}}$

We are now ready to determine the between and within mean-square (MS) values. The *mean-square (MS) values*, in a limited sense, represent the average amount of variability between and within the groups. The *mean-square-between (MSB)* value is determined by taking the sum-of-squares between value and dividing it by the degrees of freedom between. In other words, we are taking Step 9 (645) and dividing it by Step 12 (2). This gives us the MSB (reported below and in the table) of 322.5.

Step 15: The mean square (MS) between $= \dfrac{\text{SS between}}{\text{df between}} = \dfrac{\text{Step 9}}{\text{Step 12}}$

$$= \frac{645}{2} = \boxed{322.5}$$

Similarly, to obtain the *mean square within* (MSW—Step 16), we simply divide the sum-of-squares within value by the within degrees of freedom. That is, we divide Step 10 (244.5) by Step 13 (27 df), which equals $9.05\overline{5}$.

Step 16: The mean square (MS) within $= \dfrac{\text{SS within}}{\text{df within}} = \dfrac{\text{Step 10}}{\text{Step 13}}$

$$= \frac{244.5}{27} = \boxed{9.055\bar{5}}$$

Step 17 gives us the *F-obtained* value. As already noted, *F*-obtained values are simply ratio values that compare the amount of variability found between versus within groups. As such, this calculation requires that we divide the mean square between (MSB) by the mean square within (MSW); this gives our *F*-obtained ratio value. For the present calculations, we take Step 15 (322.5) and divide it by Step 16 (9.0555) to give us an *F* obtained of 35.6137.

Step 17: *F* obtained $= \dfrac{\text{MS between}}{\text{MS within}} = \dfrac{\text{Step 15}}{\text{Step 16}} = \dfrac{322.5}{9.0555} = \boxed{35.6137}$

As with previous obtained values from various other statistical tests we have explored in this text, we are now required to compare the above *F*-obtained value to an *F-critical* value to determine if there is statistical significance. Critical values of *F*, once again, are found in Appendix 3. Listed at the top of each of the tables in Appendix 3 is an explanation that first-row values are for the .05 alpha level whereas second-row values are for the .01 alpha level. Within the actual tables are column headings for the numerator degrees of freedom and row headings for the denominator degrees of freedom. These represent values that have already been calculated and are found in our ANOVA Summary Table. Numerator degrees of freedom are between df while denominator degrees of freedom are within df. In our present analysis, looking at the ANOVA Summary Table (Table 11.3), we already know that there are 2 between df (numerator degrees of freedom) and 27 within df (denominator degrees of freedom). Using these values we simply look to the appendix and find that the *F*-critical value at alpha level .05 is 3.35 and at .01 it is 5.49.

The final determination we have to make is if there is a statistically significant difference. As with Z tests and *t* tests, if the obtained value is larger than the critical value, we conclude that there is statistical significance. In the present case, since the *F*-obtained value of 35.6137 is larger than 3.35 or 5.49, it is considered significant at both of these levels. Since significance at an alpha of .01, however, is more desired than significance at the .05 level, we report our *F* obtained significant at this more desirable level.

Differing somewhat from how we previously denoted significance, on an ANOVA Summary Table we report it significant using a "$p < .05$" or "$p < .01$" representation. Since the obtained value of 35.6139 is significant at the .01 alpha level, it is reported as such with a $p < .01$. What $p < .01$ tells us is that not only is our obtained value significant at the .01 level but the probability that a Type I Error has been committed is less than 1% (.01 chance). This step, 18, and the format in which it is reported in the ANOVA Summary Table are found below.

TABLE 11.3 ANOVA Summary Table for Duck Data

Source	*SS*	*df*	*MS*	*F*	*p*
Between	645	2	322.5	35.6137	$p < .01$
Within	244.5	27	9.0555		
TOTAL	889.5	29			

Step 18: Determining F-critical value = F crit. (determined by alpha level and numerator and denominator = degrees of freedom)

$$F \text{ crit.}_{27\,(\text{denominator df})}^{2\,(\text{numerator df})} = \text{alpha } .05 = F \text{ crit.}_{27}^{2} = \boxed{3.35} \text{ and}$$

$$\text{alpha } .01 = F \text{ crit.}_{27}^{2} = \boxed{5.49} \text{ with } 35.6139 > 5.49\text{; thus, it is}$$

reported in the table as $\boxed{p < .01}$ and statistically significant.

As a hypothetical aside, if the F-obtained value was 4 in the present example, how would we report this? Since this value is larger than 3.35 but less than 5.49, we would report this F obtained to be significant at the .05 level; $p < .05$. Alternatively, if we had obtained an F-obtained value of 3, we would leave the table blank or simply write "not significant."

While an F obtained of 35.6137 clearly informs us that one or more of the means are significantly different, as noted at the beginning of this chapter, we are not told which of the groups are actually different. Stated in the terms of the data set, while we are

TABLE 11.4 18 Steps for ANOVA Summarized

Steps			
1.	−135	10.	244.5
2.	1497	11.	889.5
3.	−4.5	12.	2
4.	1210	13.	27
5.	40	14.	29
6.	2.5	15.	322.5
7.	1252.5	16.	9.0555
8.	607.5	17.	35.6137
9.	645	18.	$p < .01$

sure that one or more of the treatments of acid and vinegar is causing the ducks to lose a significant number of feathers, we are not sure if this is in comparison to the control group, each other, or both. Although there is an array of appropriate statistical techniques that can be used to determine which of the groups are actually different, we now explore one of the simplest and widely used of these tests: an LSD *t* test.

LEAST-SIGNIFICANT-DIFFERENCE (LSD) *t* TEST

To use an LSD test or an array of similar statistical tests, we must first obtain an *F* value that is significant at least at the .05 alpha level. (Although some researchers use an alpha level of .10, as previously noted, we always use the more common .05 alpha level as our minimum.) In the present analysis we are undertaking, not only is an *F* obtained of 35.6137 significant at the .05 level but it is also significant far beyond the .01 level. As such, it is obviously appropriate for us to undertake subsequent tests to determine definitively which of the means is statistically different.

The actual formula used to make this determination, found below, should look quite familiar to you. In fact, the formula for the LSD *t* test is nearly identical to a previous test we explored in Chapter 9: an Independent-Samples *t* test. The only real difference between the two formulas is that we already know what the s^2 pooled value is using an LSD *t* test whereas for the aforementioned Independent Samples *t* test we had to calculate this value. As noted below, with an LSD *t* test we simply use the mean square within (MSW) value found in our previously constructed ANOVA Summary Table as the s^2 pooled value.

Using the MSW value in our calculations has one important advantage beyond the obvious that we do not need to calculate this value separately. Since the MSW value is derived from the within df, the df used to determine the *t*-critical value (Appendix 2) are the same as the ANOVA within df. (Recall that the more df we have, the easier it is to find statistical significance.) For the present example, even though in any given LSD *t* test a group of 10 is compared to another group of 10 (20 total), since the MSW value was determined using 27 within df, we use the more advantageous value of 27 df. Thus, offering a preset alpha level of .01 with 27 df, the corresponding *t*-critical value for the following *t* tests is 2.771.

$$\text{LSD } t \text{ test} = \frac{\bar{x}_i - \bar{x}_j}{\sqrt{\text{MSW}\,(1/n_i + 1/n_j)}}$$

The reason an *i* and *j* subscript are listed on the means and the "*n*'s" of this formula is that when we are dealing with more than two means, there are obviously more than two groups to compare. Thus, we are instructed to undertake all possible tests, that is, *i* to *j* or from beginning to end. In the present example of three group means, there are three correspondingly *t* tests we can undertake: group 1 versus group 2, group 1 versus group 3, and group 2 versus group 3. To this end, the appropriate calculations for these three LSD *t* tests are completed below.

Please note the following four provisions as you review the calculations for an LSD *t* test. (1) As the formula requires, we must calculate a separate mean value for each of the groups. (2) The MSW value, once again, is taken off the previously constructed ANOVA Summary Table; in this case it is 9.0555. (3) In the actual calculations, algebraically when you minus a minus—subtract a negative value from another negative value—this is the same as adding it. Thus $-11 - (-2) = -9$; if this is still confusing, think of it in terms of integers. (4) In cases such as the present one where each of the group sizes is equal (10), once we have completed the denominator portion of the formula for the first calculation (1.3458), we do not need to recalculate it for subsequent tests.

$$\bar{x} = (\Sigma X)/n \quad \text{or} \quad \bar{x}_1 = -110/10 = \boxed{-11}$$

$$\bar{x}_2 = -20/10 = \boxed{-2}$$

$$\bar{x}_3 = 5/10 = \boxed{-.5}$$

Step 1. The LSD t test for group 1 versus group 2 is:

$$\frac{-11 - (-2)}{\sqrt{(9.0555)\,(.20)}} = \frac{-9}{1.3457} = \boxed{-6.6879}$$

t obtained $= |-6.6879| > 2.771 = t$ critical. Thus, group 1 is statistically different from group 2 at the .01 alpha level.

Step 2. The LSD t test for group 1 versus group 3 is:

$$\frac{-11 - (-.5)}{\sqrt{(9.0555)\,(.20)}} = \frac{-10.5}{1.3457} = \boxed{-7.8026}$$

t obtained $= |-7.8026| > 2.771 = t$ critical. Thus, group 1 is statistically different from group 3 at the .01 alpha level.

Step 3. The LSD t test for group 2 versus group 3 is:

$$\frac{-2 - (-.5)}{\sqrt{(9.0555)\,(.20)}} = \frac{-1.5}{1.3457} = \boxed{-1.1146}$$

t obtained $= |-1.1146| < 2.771 = t$ critical. Thus, group 2 is NOT statistically different from group 3 at the .01 alpha level.

As the above indicates, while group 2 is not statistically different from group 3, we are able to conclude that group 1 is significantly different from both groups 2 and 3. So, in reference to the original data set, what does this mean? Quite simply, the acid is causing the ducks in this treatment group to lose a statistically significant number of feathers in comparison to the ducks in the other treatment group and the control group. Vinegar, on the other hand, with this sample size does not appear to bring about a significant loss of feathers in comparison to the control group. Thus, if we are a duck concerned with keeping our feathers we should obviously avoid acid being poured off our backs!

Analyses Using ANOVA Summary Tables

An alternative approach to undertaking ANOVA, as previously noted, is to practice doing this type of analysis using just ANOVA Summary Tables. Approaching ANOVA in this manner is less time consuming because it entails far fewer calculations—eight steps—than our previous use of actual raw data and its accompanying 18 steps. Further (like most statistical techniques), the more you do the actual ANOVA calculations, the more likely you are to understand what the statistical technique actually measures. We believe, however, that the last eight steps and what they measure are the most critical in accomplishing this learning goal. As a result, we review this alternative way of approaching ANOVA in detail and offer several practice examples in the Practice Excercises of this chapter.

CARTOON 11.2

FRANK AND ERNEST reprinted by permission of UFS, Inc.

With this in mind, we introduce Cartoon 11.2 . Let's use our current research interests to see how successful different types of sex education classes are. To this end, we draw a random sample of 48 research subjects and randomly assign them to four groups of 12. Three of the groups (the treatment groups) reflect different approaches to teaching sex education, whereas the members of the control group receive no instruction. A test is administered at the end of semester with results that yield a sum of squares within value of 220 and a TOTAL SS value of 265; we do not know what the SS-between value is. This limited information is found in an incomplete ANOVA Summary Table (Table 11.5). Our task is to fill in all of the missing values.

TABLE 11.5 Incomplete ANOVA Summary Table for Sex Education Data ($J = 4$; $N = 48$)

Source	*SS*	*df*	*MS*	*F*	*p*
Between					
Within	220				
TOTAL	265				

Although there are several pieces of information that most of you could fill in immediately, we believe it is important discuss the table as an incremental, step-by-step procedure. As such, the first thing we will determine is the missing SS-between value. To do this, we simply subtract the SS-within from the TOTAL- SS to determine the missing value; in this case, 265 − 220 = 45 SS between. Conversely, if only the SS-between and TOTAL-SS values are given, we subtract the SS-between from the TOTAL to get the missing value. Finally, if just the SS-between and SS-within are listed, we simply add these two values together to get the TOTAL-SS (for this example, 45 + 220 = 265).

While we discuss the remaining steps in the terms of Steps 12–18 presented in the original ANOVA, the remaining steps should start to come somewhat automatically with practice. All of these steps are now briefly outlined and summarized in a completed ANOVA Summary Table (Table 11.6).

Step 12, 13, and 14 are used to ascertain the degrees of freedom found in three different applications. Step 12 requires a determination of the between df; $J - 1 =$ between df, or (for the present analysis) $4 - 1 = 3$ between df. The next step, 13, informs us how many within df are present; $N - J =$ within df, or (for the present example) $48 - 4 = 44$ within df. Finally, Step 14 asks for the TOTAL df; $N - 1 =$ TOTAL df, or $48 - 1 = 47$ TOTAL df.

Steps 15 and 16 ask us to determine two different mean-square values. Specifically, Step 15 is the SS-between value divided by the df between, or (for this example) the MS-between = 45/3 = 15. Similarly, Step 16 is the SS-within divided by the df within, or (for this example) the MS-within = 220/44 = 5.

The remaining steps give us an *F*-obtained value (Step 17) which is then compared to an *F*-critical value to determine if there is statistical significance (Step 18). The *F*-obtained value is derived by dividing the MSB by the MSW; or 15/5 = an *F*-obtained value of 3. *F*-critical values are found in Appendix 3, part of Step 18, and are determined by numerator and denominator df taken directly from the ANOVA Summary Table. In this case, we have 3 numerator df and 44 denominator df; as such, the corresponding *F*-critical values are 2.82 (alpha = .05) and 4.26 (alpha = .01). Thus, with an *F* obtained of 3, it is significant at the .05 alpha level (3 > 2.82) but it is not significant at the .01 alpha level (3 < 4.26); as such, $p < .05$ is reported in the table below.

TABLE 11.6 Completed ANOVA Summary Table for Sex Education Data ($J = 4; N = 48$)

Source	*SS*	*df*	*MS*	*F*	*p*
Between	45	3	15	3	$p < .05$
Within	220	44	5		
TOTAL	265	47			

Reflecting on our ANOVA Summary Table, we can also conclude that one or more of our treatment groups and/or control group is significantly different from one or more of the other(s). Since we do not know the sum of X totals (e.g., $\Sigma X.1$) so that group means could be determined for each of the groups, we cannot do the subsequent LSD tests.

Turning to our next example, which is also based upon the sex education cartoon (11.2), we now find ourselves with three groups of 10, i.e., 30 randomly placed research subjects. Our research question, similarly, is whether different types of sex education classes are more successful than others. The population from which our sample is drawn is 9th grade students at a local Junior High School. The subjects in group 1 receive traditional sex education instruction for the period of one semester, and then are tested. The subjects in group 2 receive traditional sex education and stork theory instruction for the same period of one semester, and then are tested. Members of group 3 were given no formal sex education instruction; they are simply tested (our control group). The actual test is a 10-item instrument with scores ranging from 10 (100% correct) to 0 (0% correct).

As found in Table 11.7, we know what the SS-between and the TOTAL-SS values are for this example. Additionally, for this example, we also know what the sum of X totals are for each group; $\Sigma X.1 = 85$, $\Sigma X.2 = 35$, $\Sigma X.3 = 15$. If appropriate, our F obtained is significant; knowing these values will enable us to complete subsequent LSD t tests.

Since this is at least the third time you have been through these operations, we have dropped the step-by-step format; however, should any of the current presentation become confusing, please refer back to one of the previous examples. All of the operations are also found in a completed ANOVA Summary Table (Table 11.8). Please note that we have also placed three division symbols in the table where these operations are appropriately undertaken.

First we must determine the missing SS-within value, accomplished by subtracting the given SS-between value from the TOTAL SS, or $313.5 - 260 = 53.5$. Next we determine the df between ($J - 1$, or $3 - 1 = 2$ df between), df within ($N - J$, or $30 - 3 = 27$ df within), and TOTAL df ($N - 1$, or $30 - 1 = 29$ TOTAL df). Our derived between and within df values are then divided into their corresponding SS values to give us the MSB and MSW values: MS between = SS between/between df, or $260/2 = 130$; and MS within = SS within/within df, or $53.5/27 = 1.9814$. To get the F-

TABLE 11.7 Incomplete ANOVA Summary Table for Sex Education Data ($J = 3$; $N = 30$)

Source	*SS*	*df*	*MS*	*F*	*p*
Between	260				
Within					
TOTAL	313.5				

TABLE 11.8 Completed ANOVA Summary Table for Sex Education Data ($J = 3; N = 30$)

Source	*SS*	*df*	*MS*	*F*	*p*
Between	260 /	2	130	65.6101	$p < .01$
Within	53.5 /	27	1.9814		
TOTAL	313.5	29			

obtained ratio value, we simply divide the MSB value by the MSW value: F obtained = MSB/MSW, or 130/1.9814 = 65.6101. The F-obtained value is then compared to two F-critical values which are determined by using 2 numerator and 27 denominator df: alpha = .05, F critical = 3.35; alpha = .01, F critical = 5.49. Since the F-obtained value is larger than both of these critical values, we report our answer statistically significant at the .01 alpha level; $p < .01$.

While we now know that one or more of the groups, in terms of scores on our 10-point test, are significantly different, we do not know which one(s). To make this determination, once again, we use an LSD t test. Of course, this is now possible since we know what the sum of X totals are for each group; $\Sigma X.1 = 85$, $\Sigma X.2 = 35$, $\Sigma X.3 = 15$. To assist in our calculations, we have replicated the LSD t-test formula below.

$$\text{LSD } t \text{ Test} = \frac{\bar{x}_i - \bar{x}_j}{\sqrt{\text{MSW}\,(1/n_i + 1/n_j)}}$$

Obviously, as the formula asks, we must first calculate mean values for each of the groups. The three mean values and the rest of the t-test calculations are all found below. We have preset the alpha levels at .05 with 27 df; thus, the t critical = 2.052.

$$\bar{x} = (\Sigma X)/n \quad \text{or} \quad \bar{x}_1 = 85/10 = \boxed{8.5}$$

$$\bar{x}_2 = 35/10 = \boxed{3.5}$$

$$\bar{x}_3 = 15/10 = \boxed{1.5}$$

Step 1. The LSD t test for group 1 versus group 2 is:

$$\frac{8.5 - 3.5}{\sqrt{(1.9814)\,(.20)}} = \frac{5}{.6295} = \boxed{7.9428}$$

t obtained = 7.9428 > 2.052 = t critical. Thus, group 1 is statistically different from group 2 at the .05 alpha level.

Step 2. The LSD t test for group 1 versus group 3 is:

$$\frac{8.5 - 1.5}{\sqrt{(1.9814)\,(.20)}} = \frac{7}{.6295} = \boxed{11.1199}$$

t obtained $= 11.1199 > 2.052 = t$ critical. Thus, group 1 is statistically different from group 3 at the .05 alpha level.

Step 3. The LSD t test for group 2 versus group 3 is:

$$\frac{3.5 - 1.5}{\sqrt{(1.9814)(.20)}} = \frac{2}{.6295} = \boxed{3.1771}$$

t obtained $= 3.1771 > 2.052 = t$ critical. Thus, group 2 is statistically different from group 3 at the .05 alpha level.

Summarizing our calculations, we find that the 10 subjects that receive just traditional sex education instruction answered an average of 8.5 (or 85%) of the questions correctly. This is very much in contrast to the 3.5 (35%) average of the other treatment group and the 1.5 (15%) average of the control group. Not surprisingly, our LSD t tests support this apparent difference; the scores for those that receive just traditional sex education (group 1) instruction are statistically different from those in groups 2 and 3. Further, those in the second treatment group who receive instruction in both traditional sex education and stork-theory teachings are found to score significantly higher than those in the control group.

Interpreting these findings in the context of the actual data set allows us to conclude that while a mixture both traditional sex education and stork-theory teachings (group 2) does bring about a significant increase in the number of questions answered correctly, such an approach makes no sense in light of the 85% correct response found for group 1. In other words, while the mixed approach of group 2 is better than nothing (group 3), instructing the students with just a traditional sex education emphasis (group 1) leads to the most informed students. Perhaps the disgruntled student in the cartoon can get stork-theory teachings elsewhere and where most of us do growing up: from our parents and peers.

Chapter Summary and Conclusions

This chapter explored the statistical technique of ANOVA. This technique was created and is largely used in situations where three or more group means are to be compared. ANOVA, quite simply, allows us to make an initial determination of whether subsequent tests are to be undertaken. That is, while ANOVA tells us if one or more of the group means is significantly different from one or more of the other group means, this technique does not tell us which of the actual means may be different. To make this determination, and only when a significant F is obtained, is an array of different statistical tests that determine differences between group means. For our purposes, we explored a technique called an LSD t test. This type of test, very similar to the Independent-Samples t test presented in Chapter 9, enabled us to determine which of the group means are or are not different from each other, and to make specific conclusions about the data sets from which they are calculated.

Key Terms To Remember

Analysis of Variance (ANOVA)

Sum of Squares (SS) Between	Mean Square Between (MSB)
Sum of Squares (SS) Within	Mean Square Within (MSW)
TOTAL Sum of Squares (SS)	*F* Obtained
Between Degrees of Freedom (df)	*F* Critical
Within Degrees of Freedom (df)	LSD *t* Test
TOTAL Degrees of Freedom (df)	ANOVA Summary Table

Practice Exercises

1. Using Cartoon 9.2 as a backdrop for the data found in Table 11.9, let's say our current research interests are if different types of sex education have an effect on correct responses given on a sex education test (20 questions). The population we sample

TABLE 11.9 Sex Education Data

One-way ANOVA for Number of Sex Ed. Test Questions Correct

	Group 1			*Group 2*			*Group 3*	
s	*X.1*	*X.1*^{2}	*s*	*X.2*	*X.2*^{2}	*s*	*X.3*	*X.3*^{2}
X11	18		X12	12		X13	7	
X21	14		X22	10		X23	6	
X31	11		X32	8		X33	7	
X41	13		X42	8		X43	8	
X51	14		X52	11		X53	5	
X61	16		X62	9		X63	6	
X71	20		X72	8		X73	7	
X81	14		X82	9		X83	9	
Σ*X*.1			Σ*X*.2			Σ*X*.3		
Σ*X*.1^2			Σ*X*.2^2			Σ*X*.3^2		

($n_1 = 8$) ($n_2 = 8$) ($n_3 = 8$)
(All for randomly selected respondents given a sex education test)
Group 1: Traditional sex education.
Group 2: Traditional sex education and stork-theory teachings.
Group 3: Control, given no treatment (no sex education).

three groups of eight from are 9th grade students ($N = 24$). Respondents in group 1 are given traditional sex education, and then tested. Respondents in group 2 are given traditional sex education combined with stork-theory teachings, and then tested. Finally, respondents in group 3 are considered our control group—no treatment was given, and they are simply tested.

a. Using the 18 steps for undertaking ANOVA, construct an ANOVA Summary Table for this data set.

b. Is there a significant difference between two (or more) of the groups? Why?

2. Upon completion of the above, calculate the *t*-test values comparing each group (3 *t*-tests). Also, indicate whether the obtained *t* values are significant at the 99% confidence level (alpha $= .01$).

3. Table 11.10 is an ANOVA Summary Table for two groups with 20 respondents in each ($N = 40$).
Fill in all of the missing values. (Remember: Although ANOVA was created and is most often utilized in situations where three or more group means are present, it can be used to compare just two groups.)

TABLE 11.10 ANOVA Summary Table for Two Groups of 20

Source	*SS*	*df*	*MS*	*F*	*p*
Between					
Within	380				
TOTAL	410				

4. The following is an ANOVA Summary Table (Table 11.11) for four groups with 10 respondents in each ($N = 40$).
Fill in all of the missing values.

TABLE 11.11 ANOVA Summary Table for Four Groups of 10

Source	*SS*	*df*	*MS*	*F*	*p*
Between	4,800				
Within					
TOTAL	19,200				

5. The following is an ANOVA table (Table 11.12) for three groups with 11 respondents in each ($N = 33$).
Fill in all of the missing values.

TABLE 11.12 ANOVA Summary Table for Three Groups of 11

Source	*SS*	*df*	*MS*	*F*	*p*
Between					
Within	90				
TOTAL	132				

6. In Practice Exercise 5, let's say we are also given the following information:

$\bar{x}_1 = 22$; $\bar{x}_2 = 16.5$; $\bar{x}_3 = 11$

Using this combined information and the appropriate test (alpha = .05), is there a significant difference between the groups?

Chapter 12

An Introduction to Chi-Square and Other Non-Parametric Statistics

Many of the statistical techniques discussed in this text have been based upon two basic theoretical assumptions: (1) the use of interval-ratio level data and (2) sample data drawn from a population that is normally distributed. As previously noted, the assumption of using interval-ratio data is often violated by researchers. Nevertheless, the assumption of a normal distribution is the most rudimentary of all the theoretical assumptions underlying inferential statistics that enable us to estimate population parameters.

Often, however, researchers are faced with situations where the use of sample data drawn from a normally distributed population and/or interval-ratio data is not available or feasible. Further, there are situations where a sample drawn from a normally distributed population is not even desired. As a result (as you have probably already guessed), statisticians have developed an array of different statistical techniques to deal with such situations.

This chapter explores statistical techniques that utilize ordinal or, more typically, nominal level data. Further, these techniques are considered *distribution-free* or *non-parametric* techniques. That is, since inferential statistics with their underlying assumption of a normal distribution are used to estimate population *parameters,* they are also called *parametric statistics.* Since distribution-free statistics, on the other hand, do not make assumptions of a normal distribution to estimate population parameters, they are considered non-parametric statistics.

In their order of presentation, the specific non-parametric statistical techniques this chapter explores are the *chi-square test*, the *median test*, and the *Mann-Whitney U test*. While these are some of the most commonly used non-parametric statistical techniques, especially chi-square, they are just a "sampling" of the different non-parametric procedures that are available.

Chi-Square Test

While most (if not all) of you are not familiar with the actual calculations for a chi-square test (hereafter referred to as chi-square), many of you may have discussed this technique in a research methods course or read about it in a research report. Even though the popularity of this technique in terms of its usage has lessened somewhat in recent years with the development of more powerful alternative statistical procedures, chi-square is still a widely used technique in both the behavioral and social sciences.

Interval-ratio, ordinal, and nominal levels of measurement can all be used when undertaking chi-square. If any of these levels of measurement—types of variables—are unfamiliar, please refer to Chapter 2 for a more detailed discussion. There are, however, other statistical techniques developed for interval-ratio variables; we have explored many of these in this text. Most of these techniques are parametric. And since parametric techniques are viewed as more powerful than, and superior to, other statistical procedures, it makes no sense to use lower-level, non-parametric tests with interval-ratio data. Moreover, although many of these techniques were developed for just interval-ratio data, they are often applied (as previously noted) to ordinal level data. Thus, while some researchers may stretch the usage of many parametric procedures onto ordinal level data, most will not do so with nominal level data.

As such, non-parametric statistical techniques such as chi-square are predominantly used with nominal levels of measurement, more specifically, in cases where one nominal variable is compared to another. [Recall that we can take a nominal characteristic (e.g., breed of cow) and take an interval-ratio measurement of each group (e.g., number of times the electric fence is touched) and then use a parametric technique (independent samples *t* test) to discern if there is a difference between the two groups.] Nominal variables, as discussed in Chapter 2, are qualitative characteristics that numerical values cannot be applied to in any meaningful way. Once again, variables such as gender, hair color, or your major at college are all examples of nominal characteristics.

While no meaningful values are applicable per se to nominal variables, numerical counts of rates of occurrence for different measurements of a given characteristic can be used for analysis purposes. For instance, counts of occurrence are easily applied to each of the aforementioned examples of nominal characteristics; in your statistics class you could count the number of men versus women; the number of people with blond, black, red, or brown hair; or the number of people majoring in the social versus behavioral sciences.

Gender, hair color, and major are all single counts of one nominal variable; as such, they tell us nothing about potential variable relationships. When, however, nominal characteristics are conceptualized as independent and dependent variables using counts and chi-square, we can test for statistically significant differences. In a limited sense, this is done in a manner similar to Pearson's *r*, where the expected relationship between two variables is compared to the observed one: the amount of variation in the dependent due to regression versus error. That is, nominal characteristics can also be conceptualized in terms of *expected* versus *observed* values.

Then again, since nominal variables do not involve numerous measurements of a relationship between two variables, an alternative format for measuring expected and observed has to be utilized. This is accomplished by measuring expected versus observed frequencies of occurrence in terms of proportions. Proportions, once again, are relative frequencies of occurrence of measurable events in predetermined categories. These categories, identical to those used in Chapter 3 to construct frequency distributions, must be *mutually exclusive* and *exhaustive.* In other words, any given occurrence can fit into only one category, no more, and there must be an available category for every observation.

Once measurements of nominal variables are viewed in proportional terms of mutually exclusive and exhaustive categories, a comparison can be undertaken to determine if there is a statistically significant difference between expected versus observed rates of occurrence. This is exactly what chi-square does; it determines if there is a statistically significant difference between expected versus observed relative frequencies. In other words, allowing for potential sampling error, chi-square tests to see if there is a difference between the observed and expected relative frequencies. If such a difference is found (proportionally speaking), we conclude that a statistically significant difference exists. This also allows us to conclude that one nominal characteristic is statistically associated with (typically seen as a function of) another nominal characteristic (we accept the research hypothesis). Conversely, if a proportional difference is not found, we simply conclude that the two nominal variables are not associated together (we accept the null hypothesis).

To give some concrete meaning to what has been discussed in abstract terms so far, we offer Cartoon 12.1 as a backdrop. Let's say our present interests are what sort of effect disruptive behavior, such as fighting in the car, by adolescent bovines (cows) has on trips to MacDonalds. That is, we want to know if disruptive behavior by adolescent bovines leads to trips to MacDonalds. Or do bovine parents, like many parents, use this as an idle threat to get their offspring to behave? (Alternatively, given the cultural significance of MacDonalds in our society, human parents may very well use trips to MacDonalds as a reward for good behavior.)

The actual research question we are addressing is "Does disruptive behavior by adolescent bovines lead their parents to drop them off at MacDonalds?" In other words, our independent variable is disruptive behavior and the dependent variable is a visit to MacDonalds. More specifically, the independent variable is measured in the nominal terms of whether disruptive behavior does or does not occur while the dependent variable is measured in the nominal terms of whether the parents do or do not visit MacDonalds. As such, each variable is simply operationalized as a yes or no measurement. We sample 100 bovine parents and find the data reported in Table 12.1.

RUBES By Leigh Rubin

CARTOON 12.1

"That's it! If you kids don't start behaving, I'm taking you both to McDonald's!"

RUBES by Leigh Rubin. By permission of Leigh Rubin and Creators Syndicate.

TABLE 12.1 Visited MacDonalds by Disruptive Behavior of Adolescent Bovines

		Disruptive Behavior	
		Yes	*No*
Visited MacDonalds	*Yes*	38	15
	No	7	40

Before introducing the chi-square test formula or any calculations, it is very important to discuss the above data set in terms of what it represents and in light of the previously applied terms of proportions and observed versus expected frequencies. As already noted, chi-square is applicable to any level of measurement. As such, the technique itself can use an infinite number of categories for the independent or dependent variable. The above data set, however, is operationalized as a *two-by-two contingency table.*

Since two-by-two contingency tables are by far the most common format used for undertaking a chi-square test, it is a good idea to define clearly what they represent. The table is *two-by-two* in that it represents two groups (disruptive versus not disruptive adolescent bovines) in terms of two responses (visited or did not visit MacDonalds). *Contingency* means that one variable is seen as a determinant of another variable. Or, conversely, one variable is seen as dependent—contingent—upon the variability in another variable. In sum, while the notion of contingency is applicable to any size table for two variables, since there are only two possible ways each nominal variable can occur in this example, it is simply called a two-by-two contingency table.

The categories utilized in this two-by-two design meet the previously discussed requirement that they are mutually exclusive and exhaustive. That is, the initial research question was operationalized into four very simple response categories, and each bovine can fit into only one of these response categories. Further, for each potential response there is an available category. Thus, the categories are mutually exclusive and exhaustive.

Several different relative frequencies—proportions—can be determined from a two-by-two table. For our purposes, however, column proportions are the only ones required for chi-square calculations. Any given relative frequency is derived by taking the event of interest's rate of occurrence and dividing it by the total. So, for instance, the proportion of adolescent bovines that display disruptive behavior and subsequently visit MacDonalds is 38% (38/100 = .38).

Observed values are just that, those actually reported (e.g., 7 cows are observed as undertaking disruptive behavior and not visiting MacDonalds). Expected values are mathematically determined in a moment using the observed values. What they represent is the number of events that should be found—expected—in a given category if their rates of occurrence were entirely random. Since group sizes are almost always different, however, when expected values are calculated they also take column and row totals proportionally into account.

With the above in mind, we offer the formula below for a chi-square test. An accompanying summary table (Table 12.2) is also offered as a useful step-by-step way to complete each of the calculations and, in total, to understand what they actually measure. Further, a variant of the original table with expected values calculated in it is also found (see Table 12.3). Of course, since this is our first time through the calculations and not all of them are definitively found in the summary table (e.g., expected values), each step is discussed in the section that immediately follows it. While a more expedient chi-square formula is subsequently offered, since the following calculations give you a better idea of what is being measured and are quite easy to complete, please review them carefully.

Chi-Square Test Formula

$$\chi^2 = \sum \frac{(O - E)^2}{E}$$

Beginning with the chi-square formula, it tells us to take each O (observed) value and subtract its accompanying E (expected) value, square the resultant answer, and then divide this value by E. Once this is done for each observed value, all of these answers are totaled to give a final chi-square obtained value. The chi-square obtained is then compared to a chi-square critical value found in Appendix 4 to determine if it is significant. While Appendix 4 is discussed in detail in a moment, statistical significance for chi- square is simply determined like all previous tests; if the obtained value is larger than critical value, then the relationship is significant.

Applying this to information to the bovine data set finds that while we obviously know what the observed values are, the expected values must be mathematically determined. Expected values are derived, as reported below in the variant table of the

TABLE 12.2 Summary Table for Chi-Square Test of Adolescent Bovine Data

	Yes/Yes	*No/Yes*	*Yes/No*	*No/No*	*Total*
1. Observed (O)	38	15	7	40	100
2. Expected (E)	23.85	29.15	21.15	25.85	100
3. $O - E$	14.15	−14.15	−14.15	14.15	0
4. $(O - E)^2$	200.2225	200.2225	200.2225	200.2225	800.89
5. $\frac{(O - E)^2}{E}$	8.395	6.8686	9.4667	7.7455	32.4758

TABLE 12.3 Visited MacDonalds by Disruptive Behavior of Adolescent Bovines. Observed Values With Calculated Expected Values

		Disruptive Behavior		
		Yes	*No*	*Total*
Visited MacDonalds	*Yes*	38 (Observed) .45 × 53 = 23.85 23.85 (Expected)	15(Observed) .55 × 53 = 29.15 29.15 (Expected)	53
	No	7 (Observed) .45 × 47 = 21.15 21.15 (Expected)	40(Observed) .55 × 47 = 25.85 25.85 (Expected)	47
		45 (.45)	55 (.55)	100

original data set, by taking the column percentages and individually multiplying them by each of the row totals. Row totals (across) are simply all the values found in that row (e.g., 38 + 15 = 53) while column percentages are the column totals (top to bottom) divided by the overall sample size (e.g., 38 + 7 = 45 and 45/100 = .45). Thus, for the yes/yes category (disruptive behavior displayed and visited MacDonalds) that has an observed value of 38, its corresponding expected value is 23.85 (.45 × 53 = 23.85). This and the other three expected values are reported in Table 12.3 and in the chi-square summary table (12.2).

Having determined each of the expected values, the rest of the calculations are reported in just the summary table (Table 12.2). Here, as we previously outlined, each expected value is subtracted from its corresponding observed value (e.g., 38 − 23.85 = 14.15), squared (200.2225), and then divided by the expected value (200.2225/23.85 = 8.395). These four answers are then added together to give a chi-square obtained value of 32.4758.

Before making the determination of whether this is significant, several helpful observations about the summary table (Table 12.2) are necessary. To begin with, as an easy way to check for potential calculation errors, the sum of the expected values must always equal the sum of the observed values. Thus, in the present example, the total for both the observed and expected values is 100. Further, once each expected value is subtracted from its corresponding observed value ($O - E$), the sum of these values must equal zero. As such, this step is also an excellent double check of the calculations to this point; the sum of the observed minus expected values must equal zero for subsequent steps to be correct. Finally (as the title of the test suggests), since all answers are derived through a process of squaring, one cannot have a negative chi-square value. In other words, *all* chi- square obtained values, like *F*-obtained values, are positive values.

With this in mind, let's make the final determination of whether the obtained value is significant or not. To do this, as already noted, we need to determine a chi-square critical value. While this value is determined in a manner somewhat similar to that used in previous tests with degrees of freedom, the manner in which df are now determined is very different. Where previous tests used sample size (or some variant of it), this technique uses the number of columns and rows to decide the correct chi-square critical value. More specifically, the following formula tells us how many df are present in a given analysis.

Formula to Determine the df Present and the Chi-Square Critical Value

$(R - 1)(C - 1) = \text{df Present}$

The C represents the number columns whereas the R represents the number of rows; thus the present example (a two-by-two design) has 1 df $(2 - 1)(2 - 1) = 1 \times 1 = 1$. We then look to Appendix 4 to find that two different critical values are listed for each degree of freedom, one value for the alpha set at .05 and one value for it set at .01. The critical values for the present example (with 1 df) are 3.84 and 6.63. Since the obtained value of 32.4758 is larger than both of these values, we report significance at the .01 level; $p < .01$. In other words, we are using the same manner to report significance that was used in ANOVA Summary Tables.

Thus, hypothetically speaking, if we had obtained a chi-square value of 4 (instead of the actual 32.4758), since this value is now greater than 3.84 but smaller than 6.63, significance would be reported at the .05 level. Conversely, if a 2 had been obtained, since this value is now smaller than both critical values, we would report that there was not a significant difference.

Since statistical significance was found, what does it mean? To begin with, a significant chi-square value tells us to reject the null hypothesis and accept the research hypothesis and conclude that, proportionally speaking, there is a significant difference. In other words, more responses are being found in the yes/yes category (disruptive behavior displayed and visited MacDonalds) and in the no/no category (no disruptive behavior and did not visit MacDonalds) than we would expect if their rate of occurrence was random. Conversely, there are lower rates of occurrence in the yes/no category (no disruptive behavior and visited MacDonalds) and the no/yes category (disruptive behavior did not visit MacDonalds) than expected.

Putting this into English, we can conclude that disruptive behavior apparently leads to trips to MacDonalds. In fact, 84.44% $(38/45 = .8444)$ of the disruptive adolescent bovines visited MacDonalds. Alternatively, while no guarantee of not visiting MacDonalds, not displaying disruptive behavior appears, in comparison to disruptive behavior, to significantly reduce the probability of visiting MacDonalds. Only 27.27% $(15/55 = .2727)$ of the adolescent bovines who did not display disruptive behavior visited MacDonalds. In sum, if one is a bovine and wants to grow up and to be ground round instead of a veal cutlet, one better not display disruptive behavior.

Since the next example of a chi-square test is presented in a somewhat cursory manner, if any of the steps or derived values are confusing or unclear, please refer back to this first example. For our next example, let's say we are now interested in how the

TABLE 12.4 Visited MacDonalds by Gender of Holstein Dairy Cow

		Gender	
		Male	*Female*
Visited MacDonalds	Yes	24	16
	No	15	30

nominal characteristic of gender affects trips to MacDonalds. Further, instead of the obvious beef cows that are pictured in the cartoon, we are now specifically interested in the gender of Holstein dairy cows and its effect on trips to MacDonalds. We sample 85 Holstein cows (39 male and 46 female) at birth and then take our measurement when they are two years old to determine their status: did or did not visit MacDonalds. Our findings are reported in Table 12.4.

While just looking at the two-by-two table one can see that gender apparently has an effect on trips to MacDonalds, we can more definitively make this conclusion if statistical significance is found. To this end and as before, we offer a variant table (Table 12.5) of the original (Table 12.4) with the expected values calculated in it and a summary table (Table 12.6) of the rest of the calculations with the final answer in it.

While the calculations give us a chi-square obtained value of 6.0656, until this value is compared to a critical value we do not know if a significant difference is pre-

TABLE 12.5 Visited MacDonalds by Gender of Holstein Dairy Cow. Observed Values With Calculated Expected Values

		Gender		
		Male	*Female*	*Total*
Visited MacDonalds	Yes	24 (Observed) .4588 × 40 = 18.352 18.352 (Expected)	16 (Observed) .5412 ×40 = 21.648 21.648 (Expected)	40
	No	15 (Observed) .4588 × 45 = 20.646 20.646 (Expected)	30(Observed) .5412 × 45 = 24.354 24.354 (Expected)	45
		39 (.4588)	46 (.5412)	85

TABLE 12.6 Summary Table for Chi-Square Test for Holstein Data

	Male/Yes	*Female/Yes*	*Male/No*	*Female/No*	*Total*
1. Observed (O)	24	16	15	30	85
2. Expected (E)	18.352	21.648	20.646	24.354	85
3. $O - E$	5.648	–5.648	–5.646	5.646	0
4. $(O - E)^2$	31.8999	31.8999	31.8773	31.8773	127.5544
5. $\frac{(O - E)^2}{E}$	1.7382	1.4735	1.5439	1.3089	6.0645

sent. Since this is a two-by-two design, the critical values, once again, are 3.84 (.05) and 6.63 (.01) (recall that these two values were determined in the previous example). With the obtained value of 6.0645 being larger than 3.84 but smaller than 6.63, we report significance at the .05 alpha level; $p < .05$. Thus, we can reject the null hypothesis and accept the research hypothesis and conclude that, proportionally speaking, there is a significant difference at the .05 alpha level.

In other words, being a male Holstein cow (a bull) appears to increase the likelihood of visiting MacDonalds while being a female Holstein cow (a heifer) appears to decrease the likelihood of such visits. What sort of plausible explanations might account for this discrepancy? Are dairy farmers sexist? Well, yes. In the dairy farmers' world female cows are obviously the ones that produce milk. As such, while a few good bulls are necessary to impregnate so that the herd size is maintained and to keep the female cows lactating, beyond this, a bull's only value is how much he can fetch as a Big Mac. Of course, following this thought to its logical conclusion means that when Holsteins do have to visit MacDonalds it is usually as a young bull or an old cow. Think of it—literally billions of young male and old female Holstein cows "served."

As previously mentioned, while the above procedure for calculating chi-square gives you the best idea of what is actually being measured, there is a simpler formula that accomplishes this task more efficiently. Although this formula (below) may appear to you as quite different in comparison to the formula just used, in reality, after a few simple algebraic manipulations, both formulas are exactly same. That is, both formulas are measurements of how much difference exists between the observed versus expected values. The newly introduced formula, however, accomplishes this as one set of calculations, whereas the previous formulas accomplished this with four separate calculations added together. In other words, both formulas lead to the exact same final answer being obtained (sometimes, due to the rounding of fractions, the answers will be very close but not exactly the same). To demonstrate this, the calculations for this alternative formula are completed below using the previously analyzed Holstein data. Accompanying these calculations is a variant table of the original raw data with some of the required calculations reported in it (see Table 12.7).

TABLE 12.7 Visited MacDonalds by Gender of Holstein Dairy Cow Lettered with Column and Row Totals

		Gender				
		Male		*Female*		*Row Totals*
	Yes	*A*	24	*B*	16	*A* + *B* 24 + 16 = 40
Visited MacDonalds	No	*C*	15	*D*	30	*C* + *D* 15 + 30 = 45
	Column Totals	*A* + *C*	24 + 15 = 39	*B* + *D*	16 + 30 = 46	*N* =85

Alternative Chi-Square Test Formula

$$\chi^2 = \frac{N\,(AD - BC)^2}{(A + B)\,(C + D)\,(A + C)\,(B + D)}$$

Calculations For Alternative Chi-Square Formula Using Holstein Data

$$\chi^2 = \frac{N\,(AD - BC)^2}{(A + B)\,(C + D)\,(A + C)\,(B + D)} = \frac{85\,[(24)\,(30) - (16)\,(15)]^2}{(40)\,(45)\,(39)\,(46)}$$

$$= \frac{85\,(720 - 240)^2}{3{,}229{,}200} = \frac{19{,}584{,}000}{3{,}229{,}200} = \boxed{6.0646}$$

Since the answer is the same as obtained using the previous formula and it was discussed in detail, we will not repeat ourselves here. Nevertheless, there are still two noteworthy observations about this new formula. First, to complete the calculations the letters *A* through *D* are placed in the variant table. While the actual placement of these letters is somewhat arbitrary, for consistency purposes we suggest that you follow the above format. Second, having witnessed too many of our students make this calculation error, please don't forget to square the numerator portion of the formula. Failure to do so obviously results in an incorrect final answer.

Median Test

All of the above analyses involved relationships between two nominal characteristics. And while the most typical use of the chi-square procedure is with nominal variables, this statistic (as already noted) can be used with any level of measurement. Nevertheless, sometimes when researchers are analyzing a nominal/ordinal relationship, for simplicity purposes and often because the distribution is severely skewed—not normally distributed—they will want to use a two-by-two design. Since an ordinal variable can involve several, often numerous, ranked measurements of a given characteristic, some sort of procedure to collapse it into a nominal variable must be undertaken. One specific procedure to this end that we now turn our attention to is called a *median test*.

Cartoon 12.2 is offered to assist in this discussion. This is also the final cartoon found in this text. Since this is the last chapter, and assuming your instructor has presented the course material in the same order as found in this text, upon its completion or in the near future we assume that you will be graduating. As a graduation gift to your parents you may want to share this cartoon with them as a way of saying "Thanks. This cartoon is an example of the type of textbook that your money purchased these last four or five years." On second thought, perhaps you shouldn't.

Reflecting on this cartoon, let's say our current research interests are what sort of an effect does having paid for a daughter or son's college education have on the parents' attitude about the quality of the education their child received. To this end, at a recent graduation we asked 28 parents two simple questions: (1) did they pay for their son or daughter's education—how was it financed and (2) on a 20-point scale, how happy are they about the education their child received (1—really unhappy/dissatisfied to 20—really happy/pleased). In an attempt to keep this absurd example somewhat realistic, since most parents are happy with any college their daughter or son might graduate from with honors, we limited our sample to parents with children graduating with a 2.5 to 3.0 GPA. Thus, the nominal characteristic of parent paid (or did not pay) for education is conceptualized as the independent variable while degree of satisfaction with education received (an ordinal characteristic) is seen as a dependent variable. The resultant data set, collected from 28 individual parents, is found in Table 12.8.

Looking at either groups' score range finds two highly skewed distributions. That is, while most of the scores for either of the groups are found to occur in a range of 8 to 12, both groups have a set of extreme scores. For instance, the first group of parents who paid for the education finds three parents very unhappy with the education their son or daughter received. Conversely, the second group where the parents did not pay for their son or daughter's education finds several parents (three) expressing extreme satisfaction with the education their child received. In situations where one or more of the distributions is severely skewed (here both cases are), a median test is quite appropriate.

To this end, as the title of this test suggests, a median must be calculated. As we will see in a moment, it is the determination of a median that enables us to reconceptualize this ordinal variable into a nominal variable and subsequently analyze it as a

CARTOON 12.2

two-by-two contingency table. With a median test, the actual median value is calculated using all of the observed values from both groups. To do this, we must first rank all of the scores into one frequency distribution. This is done in Table 12.9.

With this accomplished, a median can now be calculated. Referring back to Chapter 4, we use the following formula as the first step to determine the median. This answer tells us the exact point where the median is found. While all of these calcula-

TABLE 12.8 Degree of Satisfaction With Son or Daughter's Education by Type of Financing

	How Financed	
	Parent Paid	*Other Means*
	8	9
	10	20
	2	11
	10	12
Degree of	11	8
Reported	12	19
Satisfaction	1	11
with Child's	1	10
Education	11	11
	8	12
	9	13
	11	11
	8	19
		11
		9

tions are found below and tell us that the median value for this data set is 10.5, it is still a good idea to review briefly what a median is. A median, similar to a mean and a mode, is a measure of central tendency. More specifically, it gives the point where one half of the observed values are found above and where one half of the observed values are found below (i.e., once again, like the median of a road with one half on either side).

TABLE 12.9 Ranked Frequency Distribution of the Degree-of-Satisfaction Data

Score	*Absolute Frequency*	*Score*	*Absolute Frequency*
20	1	10	4
19	2	9	3
13	1	8	4
12	3	2	1
11	7	1	2

$$\text{Median Test Formula} = \frac{N+1}{2} = \frac{28+1}{2} = \frac{29}{2} = 14.5$$

Thus, we look to the 14.5th ranked observation to find it occurs between the observed values of 10 and 11. As such, we simply add these together and divide them by two to find a median value of 10.5: (10 + 11 = 21)/ 2 = 10.5. If any of this is confusing, please refer back to Chapter 4 for a more complete discussion.

$\boxed{10.5}$ = Median

Referring back to the actual test, the calculation of a median gives a point from which two nominal categories are derived: (1) above the median and (2) below the median. Reflecting these new nominal response categories and the previous categories of parent did or did not pay for education is the two-by-two contingency table found in Table 12.10. Found within each of the categories is the number of responses occurring in it. Thus, the number of responses above the median (10.5) in the category of parent paid for education is 4; four of these parents were above the median for the satisfaction level of child's education received.

With the resultant format of a two-by-two table, a chi-square test is all that is now left to do. In other words, a median test is actually a variant form of a chi-square test; the former just involves some preliminary calculations. While the initial format to determine the chi-square obtained could be used, we use instead the simpler lettered chi-square formula. To this end, we offer the intermediate steps in Table 12.11, while the formula and all of the remaining calculations are found below.

Instead of miring ourselves with the previously explored steps that were discussed in detail, we will now just make the simple determination of whether this obtained value is significant and, if so, interpret what it means. Please, however, review all of the calculations; should any of them appear at all confusing, refer back to the previous example of this type.

TABLE 12.10 Degree of Satisfaction with Son or Daughter's Education (above or below Median) by Type of Financing

	How Financed	
	Parent Paid	*Other Means*
Above the Median	4	11
Below the Median	9	4

TABLE 12.11 Degree of Satisfaction With Son or Daughter's Education (above or below Median) by Type of Financing Lettered With Column and Row Totals

	How Financed		
	Parent Paid	*Other Means*	*Row Totals*
Above the Median	A 4	B 11	A + B 4 + 11 = 15
Below the Median	C 9	D 4	C + D 9 + 4 = 13
	A + C 4 + 9 = 13	B + D 11 + 4 = 15	N = 28

Calculations for Alternative Chi-Square Formula Using Holstein Data

$$\chi^2 = \frac{N(AD - BC)^2}{(A + B)(C + D)(A + C)(B + D)} = \frac{28\,[(4)(4) - (11)(9)]^2}{(15)(13)(13)(15)}$$

$$= \frac{28\,(16 - 99)^2}{38025} = \frac{28\,(6889)}{38025} = \frac{192892}{38025} = \boxed{5.0727}$$

Since this is a two-by-two contingency table, once again we simply use the chi-square critical values of 3.84 and 6.63. Since the obtained value of 5.0727 is larger than the first critical value, it is reported significant at the .05 level; $p < .05$. Or, alternatively, we reject the null and accept the research hypothesis and conclude that, proportionately speaking, there is a statistically significant difference.

The first conclusion we can make is that most parents report an average amount of satisfaction with their child's education. Apparently, however, for a few parents who paid for the education of a child who did not do very well (graduating with a 2.5 to 3.0 GPA), there is a propensity for them to be unhappy with the education received. Conversely, not having paid for a child's education has the opposite effect for a few parents; three of these parents are quite pleased with their child's education. That is, for a few parents who did not have to pay, they are by far the most satisfied of either group.

While we can conclude that whether or not one pays for a child's education definitely appears to affect a few parents' perception about the quality of education received, any conclusion beyond this is pure speculation. For instance, we might further hypothesize that some parents who paid for their children's education appear to blame the college or university attended for their daughter or son's shortcomings. This is in direct contrast to those (few) parents who see the school quite favorably because they did not pay a dime even though their children did not do very well. Then again, we could also speculate that some parents are extremely unhappy with the quality of their son or daughter's education because quite simply they paid for it. Or, conversely, a few of the parents who did not pay may simply be proud of their child's financing their own education—and this is reflected in their perception of the school.

In other words, the manner in which this research question was addressed, using nominal variables, gives us a very limited picture of the relationship found between the two variables. While this is somewhat of a problem whenever social phenomena are quantified, this is especially true with nominal measurements. For instance, had we compared the variable of degree of satisfaction with an alternative variable of the actual number of dollars spent (an interval-ratio measurement), a more meaningful picture of the data set might emerge. Simply speculating, we might find that it is not whether one pays for education or not, but rather how much. In other words, those parents who spent the most money for their son or daughter's education might very well be the 3 that are by far the most dissatisfied. In sum, while nominal measurements are certainly not meaningless, such variables are limited (sometimes severely) in terms of what can be learned about a given variable relationship.

MANN-WHITNEY *U* TEST

The final statistical procedure explored in this text is called a *Mann-Whitney U test*. Like the previous two statistical tests presented in this chapter, this procedure is also non-parametric and widely used. This procedure is the non-parametric counterpart of a *t* test and is used in situations where assumptions about the population distribution might be violated and/or the sample size is very small. The Mann-Whitney test does not compare means or medians, but rather measures theoretically if there is a difference between the way two populations of numbers are distributed. Thus, the null hypothesis holds that one distribution is adequate to represent both distributions, whereas the research hypothesis holds that two distributions are required to represent adequately the two data sets. In other words, the null hypothesis for a Mann-Whitney test states that there is no difference between the two distributions, whereas the research hypothesis states that there is.

Drawing upon our discussion from the end of the last section, let's say that our present research interests are whether parents who paid for their son or daughter's education are more satisfied with a private or public education. While we can expect a fair amount of variability in the amount a parent pays for a given child's education (regardless of where they graduated from), let's say that tuition, room and board, and overall costs at the public university are inexpensive while just the opposite is true of the private school. That is, it does not cost very much to attend Public U, whereas to attend Private U is quite expensive. Once again, we use the 20-point scale previously utilized and administer it to the parents who paid for the education of children who graduated with 2.5 to 3.0 GPA. Our findings are found in Table 12.12.

Similar to the presentation format of a median test, we first offer each of the calculations required for a Mann-Whitney test; then we discuss what this procedure is actually measuring in terms of the data set itself. The first step is to rank all the scores from both groups. While we also did this in a sense with the median test, for a Mann-Whitney test a corresponding rank (score) is assigned to each observed value. This is done in Table 12.13.

Before proceeding, three observations are warranted. When several scores are the same—tied—their corresponding ranks are summed and then divided by the number of such observations. In other words, tied scores report the average rank of such scores. Thus, the three 20's from the data set are inferred as representing the ranks of 1, 2, and 3; however, the average of the three ranks is 2 and it is reported as such for all three scores. Be careful when you determine the next rank. Often, when one is ranking scores in this manner, there is the propensity to use the wrong next rank (e.g., using 3 instead of the correct rank of 4 in the present example). One easy way to check that you have correctly ranked all the scores: If the very last rank is any value other than that of the sample size, you have made an error. Thus, for the present example, the last rank of 25 is correctly the same value as the sample size (25).

Second, an asterisk (*), or the lack of one, is used as a way to designate which group the score comes from. This is done to assist us in the next step we must under-

TABLE 12.12 Degree of Satisfaction with Son or Daughter's Education by School of Student Whose Parent Paid Education

	Public U	*Private U*
	11	2
	19	8
	20	11
	18	12
Degree of	11	3
Reported	18	17
Satisfaction	9	11
with Child's	10	10
Education	11	11
	8	12
	20	2
	18	
	17	
	20	

TABLE 12.13 Ranked Degree of Satisfaction Data with Corresponding Numerical Rank

Score	*Rank*	*Score*	*Rank*
20	2	11	14.5
20	2	11	*14.5
20	2	11	*14.5
19	4	11	*14.5
18	6	10	18.5
18	6	10	*18.5
18	6	9	20
17	8.5	8	21.5
17	*8.5	8	*21.5
12	*10.5	3	*23
12	*10.5	2	*24.5
11	14.5	2	*24.5
11	14.5		

take (see Table 12.14), where all ranks are placed back into the appropriate group from which they were drawn.

Finally, since group sizes often differ, it is important to note that the largest group must always be designated as group 1. This is necessary for a subsequent calculation.

Table 12.14 gives us two sets of rankings. These rankings are now used to determine if there is a statistically significant difference in the way the two sets of numbers are distributed. The actual Mann-Whitney formula (and its sub-parts) used to make this determination is presented below. Also found below are the corresponding calculations for the data set currently being analyzed. As always, these steps are discussed briefly in the section that immediately follows.

Mann-Whitney Formula and its Sub-parts

$$\text{Mann-Whitney Obtained Test Value (Z)} = \frac{U_1 - U_E}{O_U}$$

where:

$$U_1 = N_1 N_2 + \frac{N_1 (N_1 + 1)}{2} - R_1$$

$$U_E = \frac{N_1 N_2}{2}$$

$$O_U = \sqrt{\frac{N_1 N_2 (N_1 + N_2 + 1)}{12}}$$

TABLE 12.14 Ranks for Degree of Satisfaction with Son/Daughter's Education by where Graduated from Data

	Public U		*Private U*	
	Score	*Rank*	*Score*	*Rank*
Degree of Reported Satisfaction Scores and Corresponding Ranks	11	14.5	2	24.5
	19	4	8	21.5
	20	2	11	14.5
	18	6	12	10.5
	11	14.5	3	23
	18	6	17	8.5
	9	20	11	14.5
	10	18.5	10	18.5
	11	14.5	11	14.5
	8	21.5	12	10.5
	20	2	2	24.5
	18	6		
	17	8.5		
	20	2		

Mann-Whitney Calculations for Degree-of-Satisfaction Data

$$U_1 = (14)(11) + \frac{14(14+1)}{2} - 140 = 154 + 105 - 140 = \boxed{119}$$

$$U_E = \frac{(14)(11)}{2} = \frac{154}{2} = \boxed{77}$$

$$O_U = \sqrt{\frac{(14)(11)(14+11+1)}{12}} = \sqrt{\frac{4004}{12}} = \boxed{18.2665}$$

$$Z = \frac{119 - 77}{18.2665} = \boxed{2.2992}$$

$\alpha = .05$, Z Critical $= 1.96$ $\quad 2.2992 > 1.96$; thus, $p < .05$

What the Mann-Whitney test does is to calculate an expected value U_E that represents a distribution adequate to represent both distributions. This value is then subtracted from an observed value U_1 that represents the actual amount of variability found in group 1 (using an alternative formula, we could use group 2). This answer is then divided by a value representing the total standardized amount of variability in both groups.

Mathematically speaking, at this point of the text we are probably correctly guessing that all of the above calculations are quite understandable. There is one value, however, that may be somewhat confusing (R_1). Actually this value is simply obtained by totaling all of the ranked scores for group 1; i.e., $14.5 + 4 + 2 + \cdots + 2 = 140$. This value notwithstanding, all of the remaining steps simply use an N_1 that represents the size of group 1 (14 in the present example), an N_2 that represents the size of group 2 (11 in the present example), or a fixed value of 1, 2, or 12. Nevertheless, please carefully review the above calculations.

The final answer for all of the above calculations gives us an obtained value of 2.2992. As with all previous tests, this obtained value is the compared to a critical value. Fortunately, in cases were both groups sizes are larger than 8 the critical values from a Z distribution are used. Thus, since the obtained value of 2.2992 is larger than 1.96 but smaller than 2.58, we conclude that there is a statistically significant difference at the .05 alpha level; $p < .05$. Further, we also reject the null hypothesis and accept the research hypothesis, and conclude that one distribution is not adequate to represent both distributions.

Considering what the raw data set looked like, a finding of significance is not that surprising. Nevertheless, there are still some more meaningful things that can be said about this specific variable relationship. More specifically, this finding tells us that parents who paid for their daughter or son's education at Public U are far happier with the education their child received than parents who paid for an education at Private U. Apparently the amount of money that parents spend for a child's education has a direct impact on how satisfied they are with the education received for average students (GPA 2.5–3.0).

One final observation about this technique is warranted. Using just the sum of the ranks from group 2, this value can be simply compared to a table value to determine significance. To give a better idea what the statistic is actually measuring, we have purposely not included this table. Moreover, we might guess that some of you have probably had enough table values for one semester or quarter.

CHAPTER SUMMARY AND CONCLUSIONS

This chapter reviewed three basic and widely used non-parametric statistical procedures. The first procedure of chi-square, although applicable to every level of measurement, was discussed in terms of which it is most typically applied: nominal characteristics using a two-by-two contingency table. Next, a similar procedure called a median test was discussed. This test involved a nominal/ordinal variable relationship collapsed into a two-by-two so that a chi-square test can be completed. Finally, a Mann-Whitney U test was discussed. This procedure determines if one or two separate distributions is necessary to represent two sets of numbers.

Since this is the last chapter in this text, both of us wish all of you "luck" in the statistical world we reside in, one that in all "probability" you now better understand. Of course, we are expressing this sentiment at the .05 alpha level.

Key Terms To Remember

Non-Parametric	Chi-Square Test
Distribution-Free	Median Test
Two-By-Two Contingency Table	Mann-Whitney *U* Test

Practice Exercises

1. Instead of dairy cows, let's say we are now interested in the gender of the cow for the Angus breed in terms of visiting MacDonalds. Our findings are reported in Table 12.15.

   a. Using the appropriate test, does gender have a significant effect on visits to MacDonalds?

TABLE 12.15 Visited MacDonalds by Gender of Angus Cow

		Gender	
		Male	*Female*
Visited MacDonalds	Yes	28	25
	No	32	31

2. Referring back to Cartoon 12.2, let's say that instead of parents and their degree of satisfaction with their sons or daughters we are now interested in these same parents' children's reported level of satisfaction. In other words, we interview the 28 children of the parents previously discussed and ask them how happy they were with the education they received. Obviously, we already know if they did or did not pay for their own education. These new findings are reported in Table 12.16.

   a. Using a median test, is there a difference between the parents' and children's scores? If so, what?

3. Finally, instead of parents' degree of satisfaction with education at a private or public university, let's say we now want to know parents' degree of satisfaction for those who paid for their son or daughter to attend a community college versus a private college. The findings are found in Table 12.17.

   a. Using a Mann-Whitney *U* test, is there a difference between the two distributions of scores? If so, what does this mean?

TABLE 12.16 Degree of Satisfaction with Education Received by Type of Financing

	How Education Financed	
	Parent Paid	*Other Means*
	8	9
	10	2
	19	11
	10	12
Degree of	11	8
Reported	12	2
Satisfaction	9	11
with	18	9
Education	11	11
By Students	8	12
	20	8
	11	11
	8	1
		11
		9

TABLE 12.17 Degree of Satisfaction with Son or Daughter's Education by School of Student Whose Parent Paid Education

	Community College	*Private U*
	11	14
	11	13
	19	8
	20	11
Degree of	18	12
Reported	11	9
Satisfaction	18	17
with Child's	9	11
Education	10	10
	11	11
	8	12
	20	15
	18	17
	18	18
	17	17
	20	

APPENDIX 1 Table of Probabilities under the Normal Curve.

1 Z	2 *Area from Mean to Z*	3 *The Big Part*	4 *The Small Part*	1 Z	2 *Area from Mean to Z*	3 *The Big Part*	4 *The Small Part*
.00	.0000	.5000	.5000	.35	.1368	.6368	.3632
.01	.0040	.5040	.4960	.36	.1406	.6406	.3594
.02	.0080	.5080	.4920	.37	.1443	.6443	.3557
.03	.0120	.5120	.4880	.38	.1480	.6480	.3520
.04	.0160	.5160	.4840	.39	.1517	.6517	.3483
.05	.0199	.5199	.4801	.40	.1554	.6554	.3446
.06	.0239	.5239	.4761	.41	.1591	.6591	.3409
.07	.0279	.5279	.4721	.42	.1628	.6628	.3372
.08	.0319	.5319	.4681	.43	.1664	.6664	.3336
.09	.0359	.5359	.4641	.44	.1700	.6700	.3300
.10	.0398	.5398	.4602	.45	.1736	.6736	.3264
.11	.0438	.5438	.4562	.46	.1772	.6772	.3228
.12	.0478	.5478	.4522	.47	.1808	.6808	.3192
.13	.0517	.5517	.4483	.48	.1844	.6844	.3156
.14	.0557	.5557	.4443	.49	.1879	.6879	.3121
.15	.0596	.5596	.4404	.50	.1915	.6915	.3085
.16	.0636	.5636	.4364	.51	.1950	.6950	.3050
.17	.0675	.5675	.4325	.52	.1985	.6985	.3015
.18	.0714	.5714	.4286	.53	.2019	.7019	.2981
.19	.0753	.5753	.4247	.54	.2054	.7054	.2946
.20	.0793	.5793	.4207	.55	.2088	.7088	.2912
.21	.0832	.5832	.4168	.56	.2123	.7123	.2877
.22	.0871	.5871	.4129	.57	.2157	.7157	.2843
.23	.0910	.5910	.4090	.58	.2190	.7190	.2810
.24	.0948	.5948	.4052	.59	.2224	.7224	.2776
.25	.0987	.5987	.4013	.60	.2257	.7257	.2743
.26	.1026	.6026	.3974	.61	.2291	.7291	.2709
.27	.1064	.6064	.3936	.62	.2324	.7324	.2676
.28	.1103	.6103	.3897	.63	.2357	.7357	.2643
.29	.1141	.6141	.3859	.64	.2389	.7389	.2611
.30	.1179	.6179	.3821	.65	.2422	.7422	.2578
.31	.1217	.6217	.3783	.66	.2454	.7454	.2546
.32	.1255	.6255	.3745	.67	.2486	.7486	.2514
.33	.1293	.6293	.3707	.68	.2517	.7517	.2483
.34	.1331	.6331	.3669	.69	.2549	.7549	.2451

APPENDIX 1 Table of Probabilities under the Normal Curve. (*continued*)

1 Z	2 *Area from Mean to Z*	3 *The Big Part*	4 *The Small Part*	1 Z	2 *Area from Mean to Z*	3 *The Big Part*	4 *The Small Part*
.70	.2580	.7580	.2420	1.05	.3531	.8531	.1469
.71	.2611	.7611	.2389	1.06	.3554	.8554	.1446
.72	.2642	.7642	.2358	1.07	.3577	.8577	.1423
.73	.2673	.7673	.2327	1.08	.3599	.8599	.1401
.74	.2704	.7704	.2296	1.09	.3621	.8621	.1379
.75	.2734	.7734	.2266	1.10	.3643	.8643	.1357
.76	.2764	.7764	.2236	1.11	.3665	.8665	.1335
.77	.2794	.7794	.2206	1.12	.3686	.8686	.1314
.78	.2823	.7823	.2177	1.13	.3708	.8708	.1292
.79	.2852	.7852	.2148	1.14	.3729	.8729	.1271
.80	.2881	.7881	.2119	1.15	.3749	.8749	.1251
.81	.2910	.7910	.2090	1.16	.3770	.8770	.1230
.82	.2939	.7939	.2061	1.17	.3790	.8790	.1210
.83	.2967	.7967	.2033	1.18	.3810	.8810	.1190
.84	.2995	.7995	.2005	1.19	.3830	.8830	.1170
.85	.3023	.8023	.1977	1.20	.3849	.8849	.1151
.86	.3051	.8051	.1949	1.21	.3869	.8869	.1131
.87	.3078	.8078	.1922	1.22	.3888	.8888	.1112
.88	.3106	.8106	.1894	1.23	.3907	.8907	.1093
.89	.3133	.8133	.1867	1.24	.3925	.8925	.1075
.90	.3159	.8159	.1841	1.25	.3944	.8944	.1056
.91	.3186	.8186	.1814	1.26	.3962	.8962	.1038
.92	.3212	.8212	.1788	1.27	.3980	.8980	.1020
.93	.3238	.8238	.1762	1.28	.3997	.8997	.1003
.94	.3264	.8264	.1736	1.29	.4015	.9015	.0985
.95	.3289	.8289	.1711	1.30	.4032	.9032	.0968
.96	.3315	.8315	.1685	1.31	.4049	.9049	.0951
.97	.3340	.8340	.1660	1.32	.4066	.9066	.0934
.98	.3365	.8365	.1635	1.33	.4082	.9082	.0918
.99	.3389	.8389	.1611	1.34	.4099	.9099	.0901
1.00	.3413	.8413	.1587	1.35	.4115	.9115	.0885
1.01	.3438	.8438	.1562	1.36	.4131	.9131	.0869
1.02	.3461	.8461	.1539	1.37	.4147	.9147	.0853
1.03	.3485	.8485	.1515	1.38	.4162	.9162	.0838
1.04	.3508	.8508	.1492	1.39	.4177	.9177	.0823

(*continued*)

APPENDIX 1 Table of Probabilities under the Normal Curve. (*continued*)

1 Z	2 *Area from Mean to Z*	3 *The Big Part*	4 *The Small Part*	1 Z	2 *Area from Mean to Z*	3 *The Big Part*	4 *The Small Part*
1.40	.4192	.9192	.0808	1.75	.4599	.9599	.0401
1.41	.4207	.9207	.0793	1.76	.4608	.9608	.0392
1.42	.4222	.9222	.0778	1.77	.4616	.9616	.0384
1.43	.4236	.9236	.0764	1.78	.4625	.9625	.0375
1.44	.4251	.9251	.0749	1.79	.4633	.9633	.0367
1.45	.4265	.9265	.0735	1.80	.4641	.9641	.0359
1.46	.4279	.9279	.0721	1.81	.4649	.9649	.0351
1.47	.4292	.9292	.0708	1.82	.4656	.9656	.0344
1.48	.4306	.9306	.0694	1.83	.4664	.9664	.0336
1.49	.4319	.9319	.0681	1.84	.4671	.9671	.0329
1.50	.4332	.9332	.0668	1.85	.4678	.9678	.0322
1.51	.4345	.9345	.0655	1.86	.4686	.9686	.0314
1.52	.4357	.9357	.0643	1.87	.4693	.9693	.0307
1.53	.4370	.9370	.0630	1.88	.4699	.9699	.0301
1.54	.4382	.9382	.0618	1.89	.4706	.9706	.0294
1.55	.4394	.9394	.0606	1.90	.4713	.9713	.0287
1.56	.4406	.9406	.0594	1.91	.4719	.9719	.0281
1.57	.4418	.9418	.0582	1.92	.4726	.9726	.0274
1.58	.4429	.9429	.0571	1.93	.4732	.9732	.0268
1.59	.4441	.9441	.0559	1.94	.4738	.9738	.0262
1.60	.4452	.9452	.0548	1.95	.4744	.9744	.0256
1.61	.4463	.9463	.0537	1.96	.4750	.9750	.0250
1.62	.4474	.9474	.0526	1.97	.4756	.9756	.0244
1.63	.4484	.9484	.0516	1.98	.4761	.9761	.0239
1.64	.4495	.9495	.0505	1.99	.4767	.9767	.0233
1.65	.4505	.9505	.0495	2.00	.4772	.9772	.0228
1.66	.4515	.9515	.0485	2.01	.4778	.9778	.0222
1.67	.4525	.9525	.0475	2.02	.4783	.9783	.0217
1.68	.4535	.9535	.0465	2.03	.4788	.9788	.0212
1.69	.4545	.9545	.0455	2.04	.4793	.9793	.0207
1.70	.4554	.9554	.0446	2.05	.4798	.9798	.0202
1.71	.4564	.9564	.0436	2.06	.4803	.9803	.0197
1.72	.4573	.9573	.0427	2.07	.4808	.9808	.0192
1.73	.4582	.9582	.0418	2.08	.4812	.9812	.0188
1.74	.4591	.9591	.0409	2.09	.4817	.9817	.0183

(*continued*)

APPENDIX 1 Table of Probabilities under the Normal Curve. (*continued*)

1 Z	2 *Area from Mean to Z*	3 *The Big Part*	4 *The Small Part*	1 Z	2 *Area from Mean to Z*	3 *The Big Part*	4 *The Small Part*
2.10	.4821	.9821	.0179	2.45	.4929	.9929	.0071
2.11	.4826	.9826	.0174	2.46	.4931	.9931	.0069
2.12	.4830	.9830	.0170	2.47	.4932	.9932	.0068
2.13	.4834	.9834	.0166	2.48	.4934	.9934	.0066
2.14	.4838	.9838	.0162	2.49	.4936	.9936	.0064
2.15	.4842	.9842	.0158	2.50	.4938	.9938	.0062
2.16	.4846	.9846	.0154	2.51	.4940	.9940	.0060
2.17	.4850	.9850	.0150	2.52	.4941	.9941	.0059
2.18	.4854	.9854	.0146	2.53	.4943	.9943	.0057
2.19	.4857	.9857	.0143	2.54	.4945	.9945	.0055
2.20	.4861	.9861	.0139	2.55	.4946	.9946	.0054
2.21	.4864	.9864	.0136	2.56	.4948	.9948	.0052
2.22	.4868	.9868	.0132	2.57	.4949	.9949	.0051
2.23	.4871	.9871	.0129	2.58	.4951	.9951	.0049
2.24	.4875	.9875	.0125	2.59	.4952	.9952	.0048
2.25	.4878	.9878	.0122	2.60	.4953	.9953	.0047
2.26	.4881	.9881	.0119	2.61	.4955	.9955	.0045
2.27	.4884	.9884	.0116	2.62	.4956	.9956	.0044
2.28	.4887	.9887	.0113	2.63	.4957	.9957	.0043
2.29	.4890	.9890	.0110	2.64	.4959	.9959	.0041
2.30	.4893	.9893	.0107	2.65	.4960	.9960	.0040
2.31	.4896	.9896	.0104	2.66	.4961	.9961	.0039
2.32	.4898	.9898	.0102	2.67	.4962	.9962	.0038
2.33	.4901	.9901	.0099	2.68	.4963	.9963	.0037
2.34	.4904	.9904	.0096	2.69	.4964	.9964	.0036
2.35	.4906	.9906	.0094	2.70	.4965	.9965	.0035
2.36	.4909	.9909	.0091	2.71	.4966	.9966	.0034
2.37	.4911	.9911	.0089	2.72	.4967	.9967	.0033
2.38	.4913	.9913	.0087	2.73	.4968	.9968	.0032
2.39	.4916	.9916	.0084	2.74	.4969	.9969	.0031
2.40	.4918	.9918	.0082	2.75	.4970	.9970	.0030
2.41	.4920	.9920	.0080	2.76	.4971	.9971	.0029
2.42	.4922	.9922	.0078	2.77	.4972	.9972	.0028
2.43	.4925	.9925	.0075	2.78	.4973	.9973	.0027
2.44	.4927	.9927	.0073	2.79	.4974	.9974	.0026

(*continued*)

APPENDIX 1 Table of Probabilities under the Normal Curve. (*continued*)

1 Z	2 *Area from Mean to Z*	3 *The Big Part*	4 *The Small Part*	1 Z	2 *Area from Mean to Z*	3 *The Big Part*	4 *The Small Part*
2.80	.4974	.9974	.0026	3.05	.4989	.9989	.0011
2.81	.4975	.9975	.0025	3.06	.4989	.9989	.0011
2.82	.4976	.9976	.0024	3.07	.4989	.9989	.0011
2.83	.4977	.9977	.0023	3.08	.4990	.9990	.0010
2.84	.4977	.9977	.0023	3.09	.4990	.9990	.0010
2.85	.4978	.9978	.0022	3.10	.4990	.9990	.0010
2.86	.4979	.9979	.0021	3.11	.4991	.9991	.0009
2.87	.4979	.9979	.0021	3.12	.4991	.9991	.0009
2.88	.4980	.9980	.0020	3.13	.4991	.9991	.0009
2.89	.4981	.9981	.0019	3.14	.4992	.9992	.0008
2.90	.4981	.9981	.0019	3.15	.4992	.9992	.0008
2.91	.4982	.9982	.0018	3.16	.4992	.9992	.0008
2.92	.4982	.9982	.0018	3.17	.4992	.9992	.0008
2.93	.4983	.9983	.0017	3.18	.4993	.9993	.0007
2.94	.4984	.9984	.0016	3.19	.4993	.9993	.0007
2.95	.4984	.9984	.0016	3.20	.4993	.9993	.0007
2.96	.4985	.9985	.0015	3.21	.4993	.9993	.0007
2.97	.4985	.9985	.0015	3.22	.4994	.9994	.0006
2.98	.4986	.9986	.0014	3.23	.4994	.9994	.0006
2.99	.4986	.9986	.0014	3.24	.4994	.9994	.0006
3.00	.4987	.9987	.0013	3.30	.4995	.9995	.0005
3.01	.4987	.9987	.0013	3.40	.4997	.9997	.0003
3.02	.4987	.9987	.0013	3.50	.4998	.9998	.0002
3.03	.4988	.9988	.0012	3.60	.4998	.9998	.0002
3.04	.4988	.9988	.0012	3.70	.4999	.9999	.0001

APPENDIX 2 Critical Values of t[a].

For any given df, the table shows the values of t corresponding to various levels of probability. Obtained t is significant at a given level if it is equal to or **greater than** *the value shown in the table.*

	Level of Significance for One-Tailed Test					
	.10	*.05*	*.025*	*.01*	*.005*	*.0005*
	Level of Significance for Two-Tailed Test					
df	*.20*	*.10*	*.05*	*.02*	*.01*	*.001*
1	3.078	6.314	12.706	31.821	63.657	636.619
2	1.886	2.920	4.303	6.965	9.925	31.598
3	1.638	2.353	3.182	4.541	5.841	12.941
4	1.533	2.132	2.776	3.747	4.604	8.610
5	1.476	2.015	2.571	3.365	4.032	6.859
6	1.440	1.943	2.447	3.143	3.707	5.959
7	1.415	1.895	2.365	2.998	3.499	5.405
8	1.397	1.860	2.306	2.896	3.355	5.041
9	1.383	1.833	2.262	2.821	3.250	4.781
10	1.372	1.812	2.228	2.764	3.169	4.587
11	1.363	1.796	2.201	2.718	3.106	4.437
12	1.356	1.782	2.179	2.681	3.055	4.318
13	1.350	1.771	2.160	2.650	3.012	4.221
14	1.345	1.761	2.145	2.624	2.977	4.140
15	1.341	1.753	2.131	2.602	2.947	4.073
16	1.337	1.746	2.120	2.583	2.921	4.015
17	1.333	1.740	2.110	2.567	2.898	3.965
18	1.330	1.734	2.101	2.552	2.878	3.922
19	1.328	1.729	2.093	2.539	2.861	3.883
20	1.325	1.725	2.086	2.528	2.845	3.850
21	1.323	1.721	2.080	2.518	2.831	3.819
22	1.321	1.717	2.074	2.508	2.819	3.792
23	1.319	1.714	2.069	2.500	2.807	3.767
24	1.318	1.711	2.064	2.492	2.797	3.745
25	1.316	1.708	2.060	2.485	2.787	3.725
26	1.315	1.706	2.056	2.479	2.779	3.707
27	1.314	1.703	2.052	2.473	2.771	3.690
28	1.313	1.701	2.048	2.467	2.763	3.674
29	1.311	1.699	2.045	2.462	2.756	3.659
30	1.310	1.697	2.042	2.457	2.750	3.646

(continued)

APPENDIX 2 Critical Values of t[a]. (*continued*)

For any given df, the table shows the values of t corresponding to various levels of probability. Obtained t is significant at a given level if it is equal to or **greater than** *the value shown in the table.*

	Level of Significance for One-Tailed Test					
	.10	**.05**	**.025**	**.01**	**.005**	**.0005**
	Level of Significance for Two-Tailed Test					
df	**.20**	**.10**	**.05**	**.02**	**.01**	**.001**
40	1.303	1.684	2.021	2.423	2.704	3.551
60	1.296	1.671	2.000	2.390	2.660	3.460
120	1.289	1.658	1.980	2.358	2.617	3.373
∞	1.282	1.645	1.960	2.326	2.576	3.291

[a] Appendix 2 is taken from Table III of Fisher and Yates, *Statistical tables for biological, agricultural and medical research*, 6th ed., published by Longman Group UK Ltd., 1974 (previously published by Oliver and Boyd, Edinburgh), and by permission of the authors and publishers.

APPENDIX 3 Critical Values of F ($\alpha = .05$ first row; $\alpha = .01$ second row).

df for denominator	*df for numerator*											
	1	*2*	*3*	*4*	*5*	*6*	*7*	*8*	*9*	*10*	*11*	*12*
1	161	200	216	225	230	234	237	239	241	242	243	244
	4,052	4,999	5,403	5,625	5,764	5,859	5,928	5,981	6,022	6,056	6,082	6,106
2	18.51	19.00	19.16	19.25	19.30	19.33	19.36	19.37	19.38	19.39	19.40	19.41
	98.49	99.00	99.17	99.25	99.30	99.33	99.34	99.36	99.38	99.40	99.41	99.42
3	10.13	9.55	9.28	9.12	9.01	8.94	8.88	8.84	8.81	8.78	8.76	8.74
	34.12	30.82	29.46	28.71	28.24	27.91	27.67	27.49	27.34	27.23	27.13	27.05
4	7.71	6.94	6.59	6.39	6.26	6.16	6.09	6.04	6.00	5.96	5.93	5.91
	21.20	18.00	16.69	15.98	15.52	15.21	14.98	14.80	14.66	14.54	14.45	14.37
5	6.61	5.79	5.41	5.19	5.05	4.95	4.88	4.82	4.78	4.74	4.70	4.68
	16.26	13.27	12.06	11.39	10.97	10.67	10.45	10.27	10.15	10.05	9.96	9.89
6	5.99	5.14	4.76	4.53	4.39	4.28	4.21	4.15	4.10	4.06	4.03	4.00
	13.74	10.92	9.78	9.15	8.75	8.47	8.26	8.10	7.98	7.87	7.79	7.72
7	5.59	4.74	4.35	4.12	3.97	3.87	3.79	3.73	3.68	3.63	3.60	3.57
	12.25	9.55	8.45	7.85	7.46	7.19	7.00	6.84	6.71	6.62	6.54	6.47
8	5.32	4.46	4.07	3.84	3.69	3.58	3.50	3.44	3.39	3.34	3.31	3.28
	11.26	8.65	7.59	7.01	6.63	6.37	6.19	6.03	5.91	5.82	5.74	5.67
9	5.12	4.26	3.86	3.63	3.48	3.37	3.29	3.23	3.18	3.13	3.10	3.07
	10.56	8.02	6.99	6.42	6.06	5.80	5.62	5.47	5.35	5.26	5.18	5.11
10	4.96	4.10	3.71	3.48	3.33	3.22	3.14	3.07	3.02	2.97	2.94	2.91
	10.04	7.56	6.55	5.99	5.64	5.39	5.21	5.06	4.95	4.85	4.78	4.71
11	4.84	3.98	3.59	3.36	3.20	3.09	3.01	2.95	2.90	2.86	2.82	2.79
	9.65	7.20	6.22	5.67	5.32	5.97	4.88	4.74	4.63	4.54	4.46	4.40
12	4.75	3.88	3.49	3.26	3.11	3.00	2.92	2.85	2.80	2.76	2.72	2.69
	9.33	6.93	5.95	5.41	5.06	4.82	4.65	4.50	4.39	4.30	4.22	4.16
13	4.67	3.80	3.41	3.18	3.02	2.92	2.84	2.77	2.72	2.67	2.63	2.60
	9.07	6.70	5.74	5.20	4.86	4.62	4.44	4.30	4.19	4.10	4.02	3.96
14	4.60	3.74	3.34	3.11	2.96	2.85	2.77	2.70	2.65	2.60	2.56	2.53
	8.86	6.51	5.56	5.03	4.69	4.46	4.28	4.14	4.03	3.94	3.86	3.80

(continued)

APPENDIX 3 Critical Values of F (α = .05 first row; α = .01 second row). (*continued*)

df for numerator

14	*16*	*20*	*24*	*30*	*40*	*50*	*75*	*100*	*200*	*500*	∞
245	246	248	249	250	251	252	253	253	254	254	254
6,142	6,169	6,208	6,234	6,258	6,286	6,302	6,323	6,334	6,352	6,361	6,366
19.42	19.43	19.44	19.45	19.46	19.47	19.47	19.48	19.49	19.49	19.50	19.50
99.43	99.44	99.45	99.46	99.47	99.48	99.48	99.49	99.49	99.40	99.50	99.50
8.71	8.69	8.66	8.64	8.62	8.60	8.58	8.57	8.56	8.54	8.54	8.53
26.92	26.83	26.69	26.60	26.50	26.41	26.35	26.27	26.23	26.18	26.14	26.12
5.87	5.84	5.80	5.77	5.74	5.71	5.70	5.68	5.66	5.65	5.64	5.63
14.24	14.15	14.02	13.93	13.83	13.74	13.69	13.61	13.57	13.52	13.48	13.46
4.64	4.60	5.46	4.53	4.50	4.46	4.44	4.42	4.40	4.38	4.37	4.36
9.77	9.68	9.55	9.47	9.38	9.29	9.24	9.17	9.13	9.07	9.04	9.02
3.96	3.92	3.87	3.84	3.81	3.77	3.75	3.72	3.71	3.69	3.68	3.67
7.60	7.52	7.39	7.31	7.23	7.14	7.09	7.02	6.99	6.94	6.90	6.88
3.52	3.49	3.44	3.41	3.38	3.34	3.32	3.29	3.28	3.25	3.24	3.23
6.35	6.27	6.15	6.07	5.98	5.90	5.85	5.78	5.75	5.70	5.67	5.65
3.23	3.20	3.15	3.12	3.08	3.05	3.03	3.00	2.98	2.96	2.94	2.93
5.56	5.48	5.36	5.28	5.20	5.11	5.06	5.00	4.96	4.91	4.88	4.86
3.02	2.98	2.93	2.90	2.86	2.82	2.80	2.77	2.76	2.73	2.72	2.71
5.00	4.92	4.80	4.73	4.64	4.56	4.51	4.45	4.41	4.36	4.33	4.31
2.86	2.82	2.77	2.74	2.70	2.67	2.64	2.61	2.59	2.56	2.55	2.54
4.60	4.52	4.41	4.33	4.25	4.17	4.12	4.05	4.01	3.96	3.93	3.91
2.74	2.70	2.65	2.61	2.57	2.53	2.50	2.47	2.45	2.42	2.41	2.40
4.29	4.21	4.10	4.02	3.94	3.86	3.80	3.74	3.70	3.66	3.62	3.60
2.64	2.60	2.54	2.50	2.46	2.42	2.40	2.36	2.35	2.32	2.31	2.30
4.05	3.98	3.86	3.78	3.70	3.61	3.56	3.49	3.46	3.41	3.38	3.36
2.55	2.51	2.46	2.42	2.38	2.34	2.32	2.28	2.26	2.24	2.22	2.21
3.85	3.78	3.67	3.59	3.51	3.42	3.37	3.30	3.27	3.21	3.18	3.16
2.48	2.44	2.39	2.35	2.31	2.27	2.24	2.21	2.19	2.16	2.14	2.13
3.70	3.62	3.51	3.43	3.34	3.26	3.21	3.14	3.11	3.06	3.02	3.00

(*continued*)

APPENDIX 3 Critical Values of F ($\alpha = .05$ first row; $\alpha = .01$ second row). (*continued*)

df for denominator	*df for numerator* 1	2	3	4	5	6	7	8	9	10	11	12
15	4.54	3.68	3.29	3.06	2.90	2.79	2.70	2.64	2.59	2.55	2.51	2.48
	8.68	6.36	5.42	4.89	4.56	4.32	4.14	4.00	3.89	3.80	3.73	3.67
16	4.49	3.63	3.24	3.01	2.85	2.74	2.66	2.59	2.54	2.49	2.45	2.42
	8.53	6.23	5.29	4.77	4.44	4.20	4.03	3.89	3.78	3.69	3.61	3.55
17	4.45	3.59	3.20	2.96	2.81	2.70	2.62	2.55	2.50	2.45	2.41	2.38
	8.40	6.11	5.18	4.67	4.34	4.10	3.93	3.79	3.68	3.59	3.52	3.45
18	4.41	3.55	3.16	2.93	2.77	2.66	2.58	2.51	2.46	2.41	2.37	2.34
	8.28	6.01	5.09	4.58	4.25	4.01	3.85	3.71	3.60	3.51	3.44	3.37
19	4.38	3.52	3.13	2.90	2.74	2.63	2.55	2.48	2.43	2.38	2.34	2.31
	8.18	5.93	5.01	4.50	4.17	3.94	3.77	3.63	3.52	3.43	3.36	3.30
20	4.35	3.49	3.10	2.87	2.71	2.60	2.52	2.45	2.40	2.35	2.31	2.28
	8.10	5.85	4.94	4.43	4.10	3.87	3.71	3.56	3.45	3.37	3.30	3.23
21	4.32	3.47	3.07	2.84	2.68	2.57	2.49	2.42	2.37	2.32	2.28	2.25
	8.02	5.78	4.87	4.37	4.04	3.81	3.65	3.51	3.40	3.31	3.24	3.17
22	4.30	3.44	3.05	2.82	2.66	2.55	2.47	2.40	2.35	2.30	2.26	2.23
	7.94	5.72	4.82	4.31	3.99	3.76	3.59	3.45	3.35	3.26	3.18	3.12
23	4.28	3.42	3.03	2.80	2.64	2.53	2.45	2.38	2.32	2.28	2.24	2.20
	7.88	5.66	4.76	4.26	3.94	3.71	3.54	3.41	3.30	3.21	3.14	3.07
24	4.26	3.40	3.01	2.78	2.62	2.51	2.43	2.36	2.30	2.26	2.22	2.18
	7.82	5.61	4.72	4.22	3.90	3.67	3.50	3.36	3.25	3.17	3.09	3.03
25	4.24	3.38	2.99	2.76	2.60	2.49	2.41	2.34	2.28	2.24	2.20	2.16
	7.77	5.57	4.68	4.18	3.86	3.63	3.46	3.32	3.21	3.13	3.05	2.99
26	4.22	3.37	2.98	2.74	2.59	2.47	2.39	2.32	2.27	2.22	2.18	2.15
	7.72	5.53	4.64	4.14	3.82	3.59	3.42	3.29	3.17	3.09	3.02	2.96
27	4.21	3.35	2.96	2.73	2.57	2.46	2.37	2.30	2.25	2.20	2.16	2.13
	7.68	5.49	4.60	4.11	3.79	3.56	3.39	3.26	3.14	3.06	2.98	2.93

(*continued*)

APPENDIX 3 Critical Values of F ($\alpha = .05$ first row; $\alpha = .01$ second row). (*continued*)

df for numerator

14	16	20	24	30	40	50	75	100	200	500	∞
2.43	2.39	2.33	2.29	2.25	2.21	2.18	2.15	2.12	2.10	2.08	2.07
3.56	3.48	3.36	3.29	3.20	3.12	3.07	3.00	2.97	2.92	2.89	2.87
2.37	2.33	2.28	2.24	2.20	2.16	2.13	2.09	2.07	2.04	2.02	2.01
3.45	3.37	3.25	3.18	3.10	3.01	2.96	2.89	2.86	2.80	2.77	2.75
2.33	2.29	2.23	2.19	2.15	2.11	2.08	2.04	2.02	1.99	1.97	1.96
3.35	3.27	3.16	3.08	3.00	2.92	2.86	2.79	2.76	2.70	2.67	2.65
2.29	2.25	2.19	2.15	2.11	2.07	2.04	2.00	1.98	1.95	1.93	1.92
3.27	3.19	3.07	3.00	2.91	3.81	2.78	2.71	2.68	2.62	2.59	2.57
2.26	2.21	2.15	2.11	2.07	2.02	2.00	1.96	1.94	1.91	1.90	1.88
3.19	3.12	3.00	2.92	2.84	2.76	2.70	2.63	2.60	2.54	2.51	2.49
2.23	2.18	2.12	2.08	2.04	1.99	1.96	1.92	1.90	1.87	1.85	1.84
3.13	3.05	2.94	2.86	2.77	2.69	2.63	2.56	2.53	2.47	2.44	2.42
2.20	2.15	2.09	2.05	2.00	1.96	1.93	1.89	1.87	1.84	1.82	1.81
3.07	2.99	2.88	2.80	2.72	2.63	2.58	2.51	2.47	2.42	2.38	2.36
2.18	2.13	2.07	2.03	1.98	1.93	1.91	1.87	1.84	1.81	1.80	1.78
3.02	2.94	2.83	2.75	2.67	2.58	2.53	2.46	2.42	2.37	2.33	2.31
2.14	2.10	2.04	2.00	1.96	1.91	1.88	1.84	1.82	1.79	1.77	1.76
2.97	2.89	2.78	2.70	2.62	2.53	2.48	2.41	2.37	2.32	2.28	2.26
2.13	2.09	2.02	1.98	1.94	1.89	1.86	1.82	1.80	1.76	1.74	1.73
2.93	2.85	2.74	2.66	2.58	2.49	2.44	2.36	2.33	2.27	2.23	2.21
2.11	2.06	2.00	1.96	1.92	1.87	1.84	1.80	1.77	1.74	1.72	1.71
2.89	2.81	2.70	2.62	2.54	2.45	2.40	2.32	2.29	2.23	2.19	2.17
2.10	2.05	1.99	1.95	1.90	1.85	1.82	1.78	1.76	1.72	1.70	1.69
2.86	2.77	2.66	2.58	2.50	2.41	2.36	2.28	2.25	2.19	2.15	2.13
2.08	2.03	1.97	1.93	1.88	1.84	1.80	1.76	1.74	1.71	1.68	1.67
2.83	2.74	2.63	2.55	2.47	2.38	2.25	2.21	2.16	2.16	2.12	2.10

(*continued*)

APPENDIX 3 Critical Values of F (α = .05 first row; α = .01 second row). (*continued*)

df for denominator	*df for numerator* 1	2	3	4	5	6	7	8	9	10	11	12
28	4.20	3.34	2.95	2.71	2.56	2.44	2.36	2.29	2.24	2.19	2.15	2.12
	7.64	5.45	4.57	4.07	3.76	3.53	3.36	3.23	3.11	3.03	2.95	2.90
29	4.18	3.33	2.93	2.70	2.54	2.43	2.35	2.28	2.22	2.18	2.14	2.10
	7.60	5.42	4.54	4.04	3.73	3.50	3.33	3.20	3.08	3.00	2.92	2.87
30	4.17	3.32	2.92	2.69	2.53	2.42	2.34	2.27	2.21	2.16	2.12	2.09
	7.56	5.39	4.51	4.02	3.70	3.47	3.30	3.17	3.06	2.98	2.90	2.84
32	4.15	3.30	2.90	2.67	2.51	2.40	2.32	2.25	2.19	2.14	2.10	2.07
	7.50	5.34	4.46	3.97	3.66	3.42	3.25	3.12	3.01	2.94	2.86	2.80
34	4.13	3.28	2.88	2.65	2.49	2.38	2.30	2.23	2.17	2.12	2.08	2.05
	7.44	5.29	4.42	3.93	3.61	3.38	3.21	3.08	2.97	2.89	2.82	2.76
36	4.11	3.26	2.86	2.63	2.48	2.36	2.28	2.21	2.15	2.10	2.06	2.03
	7.39	5.25	4.38	3.89	3.58	3.35	3.18	3.04	2.94	2.86	2.78	2.72
38	4.10	3.25	2.85	2.62	2.46	2.35	2.26	2.19	2.14	2.09	2.05	2.02
	7.35	5.21	4.34	3.86	3.54	3.32	3.15	3.02	2.91	2.82	2.75	2.69
40	4.08	3.23	2.84	2.61	2.45	2.34	2.25	2.18	2.12	2.07	2.04	2.00
	7.31	5.18	4.31	3.83	3.51	3.29	3.12	2.99	2.88	2.80	2.73	2.66
42	4.07	3.22	2.83	2.59	2.44	2.32	2.24	2.17	2.11	2.06	2.02	1.99
	7.27	5.15	4.29	3.80	3.49	3.26	3.10	2.96	2.86	2.77	2.70	2.64
44	4.06	3.21	2.82	2.58	2.43	2.31	2.23	2.16	2.10	2.05	2.01	1.98
	7.24	5.12	4.26	3.78	3.46	3.24	3.07	2.94	2.84	2.75	2.68	2.62
46	4.05	3.20	2.81	2.57	2.42	2.30	2.22	2.14	2.09	2.04	2.00	1.97
	7.21	5.10	4.24	3.76	3.44	3.22	3.05	2.92	2.82	2.73	2.66	2.60
48	4.04	3.19	2.80	2.56	2.41	2.30	2.21	2.14	2.08	2.03	1.99	1.96
	7.19	5.08	4.22	3.74	3.42	3.20	3.04	2.90	2.80	2.71	2.64	2.58
50	4.03	3.18	2.79	2.56	2.40	2.29	2.20	2.13	2.07	2.02	1.98	1.95
	7.17	5.06	4.20	3.72	3.41	3.18	3.02	2.88	2.78	2.70	2.62	2.56

(*continued*)

APPENDIX 3 Critical Values of F ($\alpha = .05$ first row; $\alpha = .01$ second row). (*continued*)

df for numerator											
14	*16*	*20*	*24*	*30*	*40*	*50*	*75*	*100*	*200*	*500*	∞
2.05	2.02	1.96	1.91	1.87	1.81	1.78	1.75	1.72	1.69	1.67	1.65
2.80	2.71	2.60	2.52	2.44	2.35	2.30	2.22	2.18	2.13	2.09	2.06
2.05	2.00	1.94	1.90	1.85	1.80	1.77	1.73	1.71	1.68	1.65	1.64
2.77	2.68	2.57	2.49	2.41	2.32	2.27	2.19	2.15	2.10	2.06	2.03
2.04	1.99	1.93	1.89	1.84	1.79	1.76	1.72	1.69	1.66	1.64	1.62
2.74	2.66	2.55	2.47	2.38	2.29	2.24	2.16	2.13	2.07	2.03	2.01
2.02	1.97	1.91	1.86	1.82	1.76	1.74	1.69	1.67	1.64	1.61	1.59
2.70	2.62	2.51	2.42	2.34	2.25	2.20	2.12	2.08	2.02	1.98	1.96
2.00	1.95	1.89	1.84	1.80	1.74	1.71	1.67	1.64	1.61	1.59	1.57
2.66	2.58	2.47	2.38	2.30	2.21	2.15	2.08	2.04	1.98	1.94	1.91
1.98	1.93	1.87	1.82	1.78	1.72	1.69	1.65	1.62	1.59	1.56	1.55
2.62	2.54	2.43	2.35	2.26	2.17	2.12	2.04	2.00	1.94	1.90	1.87
1.96	1.92	1.85	1.80	1.76	1.71	1.67	1.63	1.60	1.57	1.54	1.53
2.59	2.51	2.40	2.32	2.22	2.14	2.08	2.00	1.97	1.90	1.86	1.84
1.95	1.90	1.84	1.79	1.74	1.69	1.66	1.61	1.59	1.55	1.53	1.51
2.56	2.49	2.37	2.29	2.20	2.11	2.05	1.97	1.94	1.88	1.84	1.81
1.94	1.89	1.82	1.78	1.73	1.68	1.64	1.60	1.57	1.54	1.51	1.49
2.54	2.46	2.35	2.26	2.17	2.08	2.02	1.94	1.91	1.85	1.80	1.78
1.92	1.88	1.81	1.76	1.72	1.66	1.63	1.58	1.56	1.52	1.50	1.48
2.52	2.44	2.32	2.24	2.15	2.06	2.00	1.92	1.88	1.81	1.78	1.75
1.91	1.87	1.80	1.75	1.71	1.65	1.62	1.57	1.54	1.51	1.48	1.46
2.50	2.42	2.30	2.22	2.13	2.04	1.98	1.90	1.86	1.80	1.76	1.72
1.90	1.86	1.79	1.74	1.70	1.64	1.61	1.56	1.53	1.50	1.47	1.45
2.43	2.40	2.28	2.20	2.11	2.02	1.96	1.88	1.84	1.78	1.73	1.70
1.90	1.85	1.78	1.74	1.69	1.63	1.60	1.55	1.52	1.48	1.46	1.44
2.46	2.39	2.26	2.18	2.10	2.00	1.94	1.86	1.82	1.76	1.71	1.68

(*continued*)

APPENDIX 3 Critical Values of F (α = .05 first row; α = .01 second row). (*continued*)

df for denominator	*df for numerator* 1	2	3	4	5	6	7	8	9	10	11	12
55	4.02	3.17	2.78	2.54	2.38	2.27	2.18	2.11	2.05	2.00	1.97	1.93
	7.12	5.01	4.16	3.68	3.37	3.15	2.98	2.85	2.75	2.66	2.59	2.53
60	4.00	3.15	2.76	2.52	2.37	2.25	2.17	2.10	2.04	1.99	1.95	1.92
	7.08	4.98	4.13	3.65	3.34	3.12	2.95	2.82	2.72	2.63	2.56	2.50
65	3.99	3.14	2.75	2.51	2.36	2.24	2.15	2.08	2.02	1.98	1.94	1.90
	7.04	4.95	4.10	3.62	3.31	3.09	2.93	2.79	2.70	2.61	2.54	2.47
70	3.98	3.13	2.74	2.50	2.35	2.23	2.14	2.07	2.01	1.97	1.93	1.89
	7.01	4.92	4.08	3.60	3.29	3.07	2.91	2.77	2.67	2.59	2.51	2.45
80	3.96	3.11	2.72	2.48	2.33	2.21	2.12	2.05	1.99	1.95	1.91	1.88
	6.96	4.88	4.04	3.56	3.25	3.04	2.87	2.74	2.64	2.55	2.48	2.41
100	3.94	3.09	2.70	2.46	2.30	2.19	2.10	2.03	1.97	1.92	1.88	1.85
	6.90	4.82	3.98	3.51	3.20	2.99	2.82	2.69	2.59	2.51	2.43	2.36
125	3.92	3.07	2.68	2.44	2.29	2.17	2.08	2.01	1.95	1.90	1.86	1.83
	6.84	4.78	3.94	3.47	3.17	2.95	2.79	2.65	2.56	2.47	2.40	2.33
150	3.91	3.06	2.67	2.43	2.27	2.16	2.07	2.00	1.94	1.89	1.85	1.82
	6.81	4.75	3.91	3.44	3.14	2.92	2.76	2.62	2.53	2.44	2.37	2.30
200	3.89	3.04	2.65	2.41	2.26	2.14	2.05	1.98	1.92	1.87	1.83	1.80
	6.76	4.71	3.88	3.41	3.11	2.90	2.73	2.60	2.50	2.41	2.34	2.28
400	3.86	3.02	2.62	2.39	2.23	2.12	2.03	1.96	1.90	1.85	1.81	1.78
	6.70	4.66	3.83	3.36	3.06	2.85	2.69	2.55	2.46	2.37	2.29	2.23
1000	3.85	3.00	2.61	2.38	2.22	2.10	2.02	1.95	1.89	1.84	1.80	1.76
	6.66	4.62	3.80	3.34	3.04	2.82	2.66	2.53	2.43	2.34	2.26	2.20
∞	3.84	2.99	2.60	2.37	2.21	2.09	2.01	1.94	1.88	1.83	1.79	1.75
	6.64	4.60	3.78	3.32	3.02	2.80	2.64	2.51	2.41	2.32	2.24	2.18

(*continued*)

APPENDIX 3 Critical Values of F (α = .05 first row; α = .01 second row). (*continued*)

df for numerator

14	16	20	24	30	40	50	75	100	200	500	∞
1.88	1.83	1.76	1.72	1.67	1.61	1.58	1.52	1.50	1.46	1.43	1.41
2.43	2.35	2.23	2.15	2.06	1.96	1.90	1.82	1.78	1.71	1.66	1.64
1.86	1.81	1.75	1.70	1.65	1.59	1.56	1.50	1.48	1.44	1.41	1.39
2.40	2.32	2.20	2.12	2.03	1.93	1.87	1.79	1.74	1.68	1.63	1.60
1.85	1.80	1.73	1.68	1.63	1.57	1.54	1.49	1.46	1.42	1.39	1.37
2.37	2.30	2.18	2.09	2.00	1.90	1.84	1.76	1.71	1.64	1.60	1.56
1.84	1.79	1.72	1.67	1.62	1.56	1.53	1.47	1.45	1.40	1.37	1.35
2.35	2.28	2.15	2.07	1.98	1.88	1.82	1.74	1.69	1.62	1.56	1.53
1.82	1.77	1.70	1.65	1.60	1.54	1.51	1.45	1.42	1.38	1.35	1.32
2.32	2.24	2.11	2.03	1.94	1.84	1.78	1.70	1.65	1.57	1.52	1.49
1.79	1.75	1.68	1.63	1.57	1.51	1.48	1.42	1.39	1.34	1.30	1.28
2.26	2.19	2.06	1.98	1.89	1.79	1.73	1.64	1.59	1.51	1.46	1.43
1.77	1.72	1.65	1.60	1.55	1.49	1.45	1.39	1.36	1.31	1.27	1.25
2.23	2.15	2.03	1.94	1.85	1.75	1.68	1.59	1.54	1.46	1.40	1.37
1.76	1.71	1.64	1.59	1.54	1.47	1.44	1.37	1.34	1.29	1.25	1.22
2.20	2.12	2.00	1.91	1.83	1.72	1.66	1.56	1.51	1.43	1.37	1.33
1.74	1.69	1.62	1.57	1.52	1.45	1.42	1.35	1.32	1.26	1.22	1.19
2.17	2.09	1.97	1.88	1.79	1.69	1.62	1.53	1.48	1.39	1.33	1.28
1.72	1.67	1.60	1.54	1.49	1.42	1.38	1.32	1.28	1.22	1.16	1.13
2.12	2.04	1.92	1.84	1.74	1.64	1.57	1.47	1.42	1.32	1.24	1.19
1.70	1.65	1.58	1.53	1.47	1.41	1.36	1.30	1.26	1.19	1.13	1.08
2.09	2.01	1.89	1.81	1.71	1.61	1.54	1.44	1.38	1.28	1.19	1.11
1.69	1.64	1.57	1.52	1.46	1.40	1.35	1.28	1.24	1.17	1.11	1.00
2.07	1.99	1.87	1.79	1.69	1.59	1.52	1.41	1.36	1.25	1.15	1.00

APPENDIX 4 Critical Values of the Chi-square Distribution for $\alpha = .05$ and $\alpha = .01$

df	$\alpha = .05$	$\alpha = .01$
1	3.84	6.63
2	5.99	9.21
3	7.81	11.3
4	9.49	13.3
5	11.1	15.1
6	12.6	16.8
7	14.1	18.5
8	15.5	20.1
9	16.9	21.7
10	18.3	23.2
11	19.7	24.7
12	21.0	26.2
13	22.4	27.7
14	23.7	29.1
15	25.0	30.6
16	26.3	32.0
17	27.6	33.4
18	28.9	34.8
19	30.1	36.2
20	31.4	37.6
21	32.7	38.9
22	33.9	40.3
23	35.2	41.6
24	36.4	43.0
25	37.7	44.3
26	38.9	45.6
27	40.1	47.0
28	41.3	48.3
29	42.6	49.6
30	43.8	50.9

INDEX